Preface

What happens when technology and society is changing too fast for the companies? There are many organizations who couldn't cope up with this transformation and eventually became irrelevant in the market. For instance, Netflix grew by 360 percent over the period of 2012 to 2016 while movie rental Blockbuster's revenue dropped to almost zero.

The concept of Digital Darwinism was introduced by Brian Solis. As per Solis, we all are witnessing Digital Darwinism, the evolution of consumer behavior when society and technology evolve faster than the ability to adapt.

With increase in connectivity and the increasing number of connected devices consumers are having numerous channels and platforms to express their opinion. This connected consumerism is altering the decision making cycles posing a threat to organizations if not adapted at the right time. Many companies and businesses are still not able to figure out the real impact of this digital transformation and are struggling to adapt with it.

Even though it poses a great threat if not adapted in the right way, it also opens a plethora of opportunities for companies to find new opportunities and create long term engagements and relationship. Emerging technologies like augmented reality, artificial intelligence and blockchain is roiling the business landscape today and creating a new experience for customers. It is all about being innovative and gaining the attention of the connected customers.

In the following chapters, let's understand the significance of these emerging technology and how it complements each other to create new domains of opportunities.

Contents

Introduction

Introduction to the Trio

AI Guide

Blockchain Guide

IoT Guide

Convergence

Applications

Conclusion

OPPORTUNITY LOSE

DONT MISS YOUR CHANCE TO BOOST REVENUE
USING EMERGING TECHNOLOGIES

COACHBLINK

Published by: Coachblink Consultancy Private Limited
Website: www.coachblink.com
Email: info@coachblink.com

Unauthorized copying of this content is prohibited and is punishable under various acts.

INTRODUCTION

1. Reading Opportunity Lose: Exposure Technique

Opportunity Lose is not a typical guide which you are supposed to understand completely, and you need not remember everything mentioned here. The purpose of the book is to increase your awareness of the Emerging technologies - Blockchain, Artificial Intelligence along with internet of things which together is a potent combo, by healthy, we mean loud enough to change the world.

So you must be wondering if there is no need to study stuff in this book, how should I deal with this book?

For finding the answer to that, the understanding of how awareness can make your life better as well as how remembering and forgetting things work is inevitable. What do you feel about forgetting something? Do you think, it is your lack of attention or your lack of capacity to remember things which is the reason for your forgetfulness? We forget things intentionally! Not everything we want to do derives from our desire to accomplish the task, the desire derived from the peer pressure or the necessity to satisfy the society. When the personal desire is not strong enough in some task, we tend to forget it because actually, we don't desire well enough for it to happen. Now it does make kind of sense, doesn't it? If not convinced about that thing about how you forget stuff you studied in school and why you studied in school, was it out of a desire to know things or was it because you had no other option or because of peer pressure to perform and catch up with the rest.

The bottom-line of the remember things concept is that if exposed to something or when we are giving a fair chance to new things, we try to dig deeper when there is a spark of interest generated, but the necessary condition is exposed yourself to more things so that you could find what vibes you up the most. When you are giving yourself enough exposure, your intuition or gut feeling carries you through the rest of the process. When generated the sparks, your intuition leads you to find more information so that you can make a revolution happen.

You can keep this book in such a way that you see this frequently when it is easily accessible, and a desire to know what is happening in this sector triggers, which gives a proper chance to get yourself exposed to the concepts of artificial intelligence, the blockchain, and the internet of things. It is not necessary to read up everything in a single read, but you can give it a fair chance by keeping the book always in your reach till you are sure you have a fair chance to understand the concepts. After all, we should not let a chance to change the world to slip by.

2. Your Opportunity lose

How can you avert potential opportunity loses that could befall on you?

There are two different situations in which an opportunity could come into play. Let's see the case of investing in Bitcoins. Which category do you belong to if we are talking about investing in bitcoins?

- Invested and made a fortune out of it
- Failed to make a fortune or suffered a lose
- Optimistic and held bitcoin in high regard, you knew it would climb high but still did not invest
- Pessimistic about the success of bitcoin
- Neutral about bitcoin, doesn't matter to you
- You were not aware of what is happening around bitcoin even though you are familiar with the term
- You did not know about bitcoin

So where do you belong?

The opportunity associated with the third situation where you knew bitcoin would be successful but you did not invest in it because of some reason, this triggers a deep sense of regret and the apparent feeling of "I knew it." The second category is the majority, and it is the category that most of us belong to - the state of unawareness. This category is the last case in the list. You did not even have a chance to explore the idea because that idea did not grab your attention enough. The principal motive of this book is, we are going to minimize the last category of people by making everyone aware of this great opportunity. We don't want to miss the next Mark Zuckerberg or the next Bill Gates because this concept failed to grab his attention correctly now. This awareness is the light trigger we need to find and mine out the best among us. This book is our contribution to the emerging technology to awaken the best ones out there. If you felt you are an ordinary person, maybe this is the trigger which can make you feel superhuman by lifting your awareness levels.

If we are thinking a bit deeper, there is indeed various degrees of opportunity loses associated with every state specified in the bitcoin stance list. This book is for everyone regardless of which category you belong to; this book aims to maximize your awareness so that it minimizes opportunity loss.

When you don't know something or when you are unaware of the existence and operation of something you are potentially losing all the fortunes which you can reap from it. Maybe it is not your cup of tea, but you would never know how your interest grows into stuff unless you give it a fair chance. This book is your chance to give a fair chance to emerging technology. The school was forced at times, all the forced assignments and project which wore us out. All that was required was to give fair exposure to all the subject so that we can act upon our freedom to choose. When you are exploring the emerging technologies, it is not just a chance of building a profit reaping industries or changing your company into a money machine, and it is about how you change the world.

Emerging technologies are our best bet for creating a revolution. All of us have thought of changing the world, some of you still believe in it, and some of you might have given up on that. For those who are still on it, this opens more doors for fulfilling your destiny by changing the world and for those who gave up this is your second chance to grip back your dreams and to fill your life with hope and ambition.

Maybe you are someone who has achieved well in life and is satisfied with what you have. I'm talking about the static state of happiness, you sit and enjoy your fortune. How long do you think it will take before this feels monotonous. It will make you feel rusty and hollow inside, and once you realize that nothing is changing, you will start to feel, the world is standing still. A little bit of extra awareness is all you need to dust off that rust. It is not necessary to work hard and spend all your effort in finding what excites you. You need good enough level of awareness.

Even if you are someone who questions the existence of everything in this world, it is okay that our perception is limited to our five senses, it is okay that you do not know everything but is it okay that you are not satisfied in your domain of control and power. The last we can possess is the control over the things we can perceive. So why not give a fair chance to the things happening in this domain of perception.

Maybe you are someone who gave your best all along, you did everything you can, but the results never showed up as you expected. What went wrong there? In that case, let me tell you this "80 percent of the results come from 20 percent of the action". The chances are your efforts were falling in 80 percent of the action which produces just 20 percent results. Emerging technologies are something which in fall into that 20 percent of effective action, the fraction which could potentially revolutionize the world. This awareness for you is the redemption to make a change and to cultivate better results.

Once you are done reading this book, how can you keep yourself refreshed with the exposure?

You can keep yourself updated at the official coachblink website
www.coachblink.com

3. Emerging technologies

Preview

- Introduction to Emerging technologies
- A brief of emerging technologies - 3D Printing, Nanotechnology, Robotics, Gene Therapy, Artificial Intelligence, Distributed Ledger Technology

"Any sufficiently advanced technology is equivalent to magic."
- Arthur C. Clarke

Emerging technologies have the capability of changing the basic functionality of the world. If you think the world has become too stable with all the available resources, which is where the effect of emerging technology kicks in and the status quo is modified. Emerging technologies are responsible for shifting the right baseline along which the world functions day after day. There is a lot of chaos around emerging technologies due to lack of legitimate information and an overflow of wrong information aimed at hyping the whole situation and creating a market advantage. Companies including blockchain along with their company name to give a false impression and thus easily obtain funds is a classic example of exploiting an emerging technology hype.

The emerging technologies have a standard set of properties like a high rate of growth, a great deal of impact and a lot of confusion around it. These technologies are not necessarily an entirely new idea build from scratch, and they can be something built from the combination or improvement of existing technologies and concepts. If these technologies become successful, the early adopters will get a lot of advantage- The Beginners Advantage, which is one of main motivation driving people to be more of the emerging technology so that they can adapt it into their

Let's check out what are the emerging technologies in the current scenario:

3D Printing

When this title goes through your head, you must be imaging anything 3D can be created using this technology. This idea goes beyond that when combined with other technologies like the internet, endless possibilities. You can even send objects to a people far away as well as online purchases are being made available as 3D printouts. When you buy products online, you get it 3D printed at your 3D printing machine instead of that product being shipped all the way to your place. 3D printing is something like teleporting an object. This innovation is indeed a promising technology to look forward. There were controversies regarding 3D printed guns in 2013.

Nanotechnology

Everything around you, everything you can see is made up of atoms and molecule. Nanotechnology is the answer to what degree the objects can be manipulated, by objects I mean almost everything from clothes to large infrastructures. In this sector, the manipulation of material in the order of nanometers is used to create macro objects. It is like using nano-sized particles like Legos to build more massive objects. The fact that materials can exhibit different properties at nanoscale makes this sector more interesting. This technology has a wide range of applications due to its ability to manipulate materials like never before.

Robotics

This sector needs no introduction, we have heard about robots from our childhood. The early exposure to different animated series, stories and movies have made us well aware of the idea of a robot. Robotics which is concerned with increasing the efficiency of the robots to perform more advanced feats can make our lives much more comfortable by automating many processes, avoiding human intervention in potentially dangerous places. Robotics can lift the role of humans by removing all sorts of tedious works and just requiring humans to be in a supervising or overseeing position.

Gene Therapy

This technology is an advanced method of treating diseases. In this process, the nucleic acid is delivered in the cells of the subject as a treatment to cure disease. The Alteration of the human genome can treat dreadful genetic diseases and even cancer.

Artificial Intelligence

AI Technology is concerned with imparting machines, the thinking capability or intelligence. When machines can think more like a human, they can take better decisions which is the basic idea behind infusing intelligence into machines. We have a whole section dedicated to artificial intelligence

Distributed Ledger Technology

DLT is a newcomer considering this list of emerging technologies. DLT is primarily a technology which enables the recording of transactions in a transparent and immutable manner. This technology has a wide range of application which covers almost every industry which involves transactions and conditions based activities. Think of some industry which does not require transactions or condition based actions? Good luck finding the answer to that! That is the level to which this technology can influence the world. This technology can create a paradigm shift, and once the idea is implemented.

In this book, we are going to dive deep into the combination which is valued more than any other combination of emerging technology. The combination which can topple the basic working of the society and create a paradigm shift in every sector imaginable. Creating a change like the introduction of the internet did at that time, and that is the magical combination of Artificial Intelligence, Internet of Things and Blockchain.

When technology is budding there a few things common to them, which repeats over and over again. Whenever something of great potential rises these things accompany the process, they are not entirely agreed upon by everyone in the society, and they are more like analysis of the situation when the technology is emerging and as we know analysis and predictions are greatly depending on perspective. Let's go through those concepts before we move into the specifics of AI, IOT, and blockchain.

Summary

- Emerging technologies can change the way the world is currently working
- We are focusing on the combination of Blockchain, Artificial Intelligence, and the Internet of Things in the following chapters.

4. Hype Cycle of emerging technologies

Preview:

- Breaking down the process of hype
- Comparing the Hype cycle to the current situation

"In my opinion, right now there's way too much hype on the technologies and not enough attention to the real businesses behind them."
- Mark Cuban

It is a fact that we get attracted to the hype. Try denying it, tell yourself you are not the kind of person who falls for hype, how long did you control yourself from clicking a hyped article? It is commendable if you had enough access to the internet and you successfully abstained from hypes for a month. We would love to study and research your brain activity for developing better marketing tools if you are a hype proof person, don't forget to reach out to us if you feel you are such. Every new trend has a hype surrounded it, especially when a factor of hope is attached to the newly found idea.

Why are we talking about hope here? The basis of everything which sells around us is hope, even the trivial thing like a cap. When you are buying a cap, you are hoping to get protected from the sun, or you are hoping to look cooler. So, you are buying hope. When something that can impart hope is hyped up, it can work wonders. Let's see the Gartner's breakdown on hype.

When a new technology is emerging it is said to follow a particular path as suggested by the IT firm Gartner, this is known as Gartner's Hype Cycle. There are five stages along which the new technology passes which is by the maturity of the project, adoption by the community and the application of that technology. This cycle is following Amara's law which states:

"We tend to overestimate the effect of a technology in the short run and underestimate the effect in the long run."

When you are coming across a new technology, you can find it fitting in one of the stages suggested by Gartner. This cycle is not an accurate depiction of what every technology goes through, and there is indeed a great deal of criticism behind this hype cycle theory. Let's check out the theory because we will never know how some information is going to help us in the long run.

The five stages of the hype cycle are:

Technology trigger

The technology is in its nascent phase in this stage. However, there is enough proof of concepts to attract media. This phase is just ideation where the market importance of the technology is not understood as well as its practical application not tested yet.

The peak of Inflated Expectations

What drives and motivates the people most? What makes people go behind a new technology? The answer is mostly the fact that money is following that project. This idea can be well propagated by familiarizing the success stories of the people who invested in this idea and the opportunity loss of the people who knew it was coming and did not do anything about that. The people under the charge of this project will make full use of the crowd speculation and continue the project with full moral and something they don't act at all.

Trough of disillusionment:

This phase is a major crossroad in the newly found technologies journey. Substantial loss of interest is observed if the project doesn't deliver what it initially promised and thus crumbling down the initial promises. If the promises are delivered, and there is a proof of progress, the early adopters of the project

Slope of Enlightenment:

There were not many who understood the idea until now, but when more and more application of the technology becomes widespread, a more significant fraction invests in understanding the idea. This positive speculation can create an increase in demand for the project and thus can attract more investments. This phase doesn't mean everyone is on board, the last layer of skeptics remain who maintain their status quo.

Plateau of productivity:

This phase is when the product is ready for mainstream adoption. The applications are well defined, and the feasibility of the project more readily understood with all the progress and working applications. At this stage, if not confined to a niche, the technology experiences tremendous growth.

Let's take the example of 2017, according to Gartner's hype cycle virtual reality was on the slope of enlightenment, augmented reality in the trough of disillusionment. At the peak of inflated expectation comes to machine learning, autonomous vehicles, and nanotube electronics. The Blockchain has also just moved into the peak of inflated expectation from innovation trigger. You can find most of the crowd in the first region, the innovation trigger - Quantum computing, smart workspace, Virtual personal assistants, Smart robots, Gesture controlled devices, Brain-computer interface, and IOT platforms.

There is much criticism going around this breakdown of hype, and The critics see it just as a post-analysis of the stuff that already happened and which in no way can help us when a new technology emerges. This theory is more of speculation which is not derived from any data and any numbers supporting these breakdown, so the phases aren't justified. The terms describing the phases are exaggerated which doesn't give a clue about what the technology is going through. The description is so vague that what is to be done in a particular stage is very unclear and moreover there are no clear-cut boundaries for the mentioned stages. Whether this idea is valid or not, you now have another element in your arsenal to analyze a situation.

Summary
- According to Garter five stages of the hype cycle are Technology Trigger, Peak of inflated expectations, Trough of Disillusionment, Slope of enlightenment and plateau of productivity.
- The hype cycle is just for analyzing an emerging technology situation which has happened or is happening.
- The hype cycle cannot help in predicting the nature of emerging technologies.

5. Technology adoption lifecycle

Preview

- Dividing the adopters based on their mentality and access.

I'm interested in things that change the world, or that affect the future and wondrous, the new technology where you see it, and you're like, 'Wow, how did that even happen? How is that possible?'

-Elon Musk

With the introduction of new technology, according to technological adoption lifecycle, the acceptance follows a particular pattern depending on the mentality and the accessibility of the people to the new technology. This graph is a bell curve distribution with the innovators being the first comers and the people who are the most skeptic - The Phobics who come in the other end of the bell curve. If you consider the group of people with whom you are interacting on a daily basis, you can identify when these people will be using a new product or a new service. There is a category who use a new product as soon as it is available in the market, they continue using that product if it is delivering what it promises. These are the people who are generally very proactive and can quickly adapt to the new situation. Then there's the other extreme who are very skeptical and are open to using the new product or service when there is no other option or when everyone around them is using that. These two categories occupy both sides of the bell curves with the others exhibiting a fraction of both behaviors.

The first users of the products- The innovators are the people with a better awareness of the technology, most people with better education and the ability to be more resourceful and proactive mentality who can quickly adapt to new changes.

Next comes, the early adopters who are educated but aren't as resourceful as the early adopters. You don't require a deal of resource actually to involve in the new technology, you need the ability to find your resources.

Early majority is the net category who are open to exploring new ideas and are active in the community and can influence other people to adopt the new idea.

Late majority includes the people who are less aware and are not open to new ideas. This category is less educated or educated to function mechanically and lack the mindset to explore new ideas. These people are also less active in the community and thus are less exposed to the new ideas.

The laggards are the other extreme who are very hard to convince for the adoption of the new product or service. They are so fixated on the status quo bias, and they will be very unwilling to change from the present state of functioning. IF the new service is something which can influence their everyday routine, they will be adopting the service if they have no other option. These are the people who are least aware of the situation, and even they are exposed they will be highly skeptical about the chances of success of the newly formed idea.

This cycle was derived from the model known as the diffusion process which was published in 1957 by Everett Rogers, George Beal, and Joe Bohlen and the model was intended to influence the agriculture economics. The objective of this model was to track the purchasing pattern of the hybrid corn seeds.

Summary

- The adopters can be divided into five categories based on the technology adoption lifecycle model.
- The first users, Early adopters, Early majority, Late majority and lastly the laggards.
- This model was adopted from the adoption model proposed by Everett Rogers.

6. Diffusion of innovations

Preview

- How technology spreads?
- Different elements associated with a spreading technology.
- The nature of adopters
- Future of Diffusion

"Any sufficiently advanced technology is indistinguishable from magic."
— Arthur C. Clarke

We are dealing with emerging technologies here, one of the exciting things about emerging technologies is the way they spread and get adopted or how the emerging technology is diffused into the crowd. This idea of diffusion of innovation was proposed by Everett Rogers in his book "Diffusion of Innovations." According to Rogers, there are several elements in the diffusion process which are the innovation, channels of communication, time window and the social system. The spreading process depends on the capital which is investing in marketing the product and the magnitude to which it is accepted by the crowd. In the chapter Technology adoption lifecycle we have seen the different categories in the societies - innovators, early adopters, early majority, late majority, and phobic.

How can we measure innovation? Innovation is defined following the existing technology and resources, and this involves a broad spectrum based on how the idea is perceived by the crowd. Based on the current state of things, we judge the magnitude of a particular innovation.

The adopters of the idea can be individuals, a group of like-minded people rooting for a particular niche or organizations like educational institutions, healthcare facilities. Which can enable large-scale adoption of the idea? Even an entire country can be an adopter of the new idea. Governments can root for the emerging technologies by supporting the startups based on these techs, conducting campaigns and even adopting in government institutions.

How does this information reach people? How are people becoming aware of the existence of such an idea? The communication channels do that job. Communication channels give for accessibility to understand the project, and it familiarizes the crowd with the idea. You can think of how you came across new ideas, to understand the effectiveness of communication channels on you.

The time window is another factor; innovations are not adopted quickly, it takes time to propagate. The knowledge about the state of adoption and planning to overcome the state can accelerate the process.

The social system is the last element in this theory, and this includes the influence of external factors like marketing campaigns and how people perceive the team or organization involved in the project. The internal influences include factors like the magnitude and nature of social interaction and distance from the potential preachers of this innovation.

The adopters consider how well this innovation can perform concerning the resources which are available in the market currently. The new product can be more efficient, but it depends on an environment which supports the product. If it is compatible with the pre-existing environment, the learning curve associated with the technology and its potential for improvement and growth are also important factors associated with the innovation. Sometimes with a steep learning curve, the product can be adopted due to the dominance of other concerned factors.

Let's see more characteristics of the innovations which were not mentioned by Rogers. Spectrum to which the innovation can be applied to, riskiness associated with the adoption, the impact of integrating the innovation with the existing system/environment. The innovation which is less risky and does not pose any possible issues for integration with the pre-existing systems are more readily adopted. When innovation is causing a variation in our daily routine, it is hard to adopt since there is a greater instability involved when a routine is changed. We are inclined to protect our status quo. The relative complexity of the innovation compared to the one which it is replacing can be dealt with more efficiency if the companies involved in earlier products and the early adopters support the adoption movement.

What are the characteristics of the people involved in the adoption of the innovation?
The motivation level of the individual plays a significant role in adoption. The willingness to adjust the status quo for a more efficient lifestyle and the proactive nature to face a problem and improve the current situation is vital for adoption. If the proactive nature is not natural, the adoption will happen when it is infused

upon the target adopters by selecting proper marketing strategies. Proper motivation can significantly drive the ability to become proactive. The proper technique involves the selling hope of a better future, which is an excellent incentive for the individual to invest in the idea. Proper marketing strategies can even trigger resourcefulness by imbibing them with the right amount of motivation. In short, the motivation of the individual decides the adoption.

When it comes to an organization which is to adopt the innovation, the process gets a whole lot complicated because of the number of people involved in the process. The same motivation and the willingness to make a change also applies to the organization. When the innovation is compatible with the existing environment, the process is not complicated, but if the upgrade demands a new environment, it will overshoot the budget involved in the process (we are considering the learning curve of the people involved in the organization who are concerned with the innovation's activity). Organizations also face pressure from the community or the economy to adopt the innovation.

There are different stages involved in the decision-making process of adopting the innovation which is knowledge, persuasion, decision, implementation, and confirmation. Knowledge stage is the first exposure due to the innovation; people are not motivated enough to dwell in details in this stage. During the persuasion stage, people feel interested in the innovation and are motivated to find out details of the project. The decision stage is crucial where the people decide whether the innovation should be accepted or rejected based on the pros and cons of the innovation. The innovation is implemented, and the effectiveness of it is experienced by the adapter in the implementation stage. When the people are satisfied with the implementation in the previous stage, they confirm the adoption of the innovation and the decision is finalized. This is the Confirmation stage.

The speed at which the innovation is adopted is known as the rate of adoption which is measured regarding the time required to convert a percentage of the target audience. The speed is defined according to the category to which the people belong to. The adoption rate reaches a critical point after a period of time, and it stays constant after that. This is when the innovation can sustain itself.

When a diffusion fails, it doesn't imply that the innovation was not adopted, but it implies that the innovation lacked fundamental value to sustain itself. The diffusion can fail when it is confined to a particular category of people, and it failed to reach others, who could benefit from the innovation. When there is not enough involvement from the community which already adopted it, it can fail because people need enough testimony to carry on the newly formed trend.

It is difficult or nearly impossible to measure the factors responsible for adoption. This is due to the complex nature of the people, and we see trends getting broken every time. The trend is broken because factors which are not considering previously so this theory is more of post-event analytic and for future analysis, a combination if this theory and a great deal of creativity, as well as gut feeling, is required.

Summary
- "Diffusion of innovations" is the model suggested by Everett Rogers to understand how technologies are adopted
- Different elements in this model are innovation, channels of communication, time window and the social system.
- The elements which were not considered by Rogers are the extent of application of the technology, risk associating with the adoption and how well the current environment can support the technology.

7. Technology acceptance model

Preview:
- A model suggesting the actors which affect adoption of a technology

"It has become appallingly obvious that our technology has exceeded our humanity."
- Albert Einstein

When we are dealing with new technologies, it is always helpful to be aware of the different model which suggest how new technology is accepted by the community. When a new technology is introduced to the world, people will be exposed more and more to the technology depending on the scale of marketing of this project. When people are exposed more and more, they will leave with the necessity to decide whether they are rooting in for the projector to ignore everything related to the project. This decision opportunity pops up each exposure of the product, and the decision behavior can change over time. According to the technology acceptance model (TAM), there are two factors which play a major rule:

The perceived usefulness of the product or service. When exposed to the product, the people think how it can improve his/ her life, and it can be something which can make that person's life better on a daily basis or a long run by helping him perform better.

The perceived ease of use of the product is another factor which people think of when exposed to the new product. It is a measure of how much people believe that using this product is going to decrease the effort while performing the concerted activity, the measure of liberation from the effort.

There is various criticism accumulating around this model in spite of its frequent usage. The criticism surrounds the facts that this model lacks proper explanation, prediction capabilities or any practical use. This model was attempted to expand, and it resulted in confusion and chaos. This model is explained in a rather basic and trivial manner. The one crucial thing which is ignored in this model is the social influence on the decision making of the individual. Whenever more and more factor explaining the perception of the individual was explained, the factor of social influence was getting more and more ignored. The behavior like herd mentality and crowd speculation are often driven by the influence of the crowd and media. This is like a keyhole vision forgetting about the full picture of what is happening.

There are even theories and studies which suggests that these perceptions do not play a role in technology adoption. According to these studies the tendency of investing time in a new service increase when a bit of complexity was added to it. The effect of the addition of complexity can be seen in the case where the sales of cake mix increased when the additional step of adding the egg to the mix was introduced. We often get attached to something, when we are willing to put some effort into supporting the working of the product. This model should be used as a means of expanding our perception rather than relying solely on this model to predict a future situation.

Summary
- According to the TAM model, the acceptance depends on perceived usefulness and the perceived ease of use of the product.

8. Economic Bubble

Preview:
- Understanding the idea of the bubble
- The Dutch Tulipmania
- What causes a bubble?

"The bubble, as investing phenomenon, has been well studied ever since the 17th-century tulip bulb frenzy. Its counterpart in bear markets is not well understood."
- Kenneth Fisher

Imagine you are on your monthly shopping spree, when you walk into the store you notice that people are grouped up in a particular aisle. Out of curiosity, you go over there to check out what's happening. There you see a rack of fancy flowers, it looks cool and all but what's the reason for this crowd, you wonder? Now you check out the price of those flowers and you are surprised by seeing how exorbitantly priced. You don't understand the reason why people are going after this, but the chances are you feel an urge to own that product because of the demand that is going through. There's always a chance that you will give in to the urge to own the product or to resell these flowers which are under high demand.

This might sound like a crazy hypothetical situation, and the fun part is something of that sort happened already in the 17th century - The Dutch Tulipmania. Tulip gained its popularity in Europe due to the brilliant colors it came in; there was an exotic vibe attached to the tulip. Further increased when a virus is known as "tulip breaking virus" enhanced its beauty by breaking the color of its petals and thus producing a multicolored petal.

These tulips named fancy like an admiral, general and every other fancy name people could think of. The scarcity factor was also playing a significant role in increasing demand as tulip flowers only after a minimum of 7 years. More and more people wanted these tulips for reselling them for enormous profits, for this purpose contracts were made which are known as futures. By buying futures for tulips, you are buying them before they are even flowered. Futures is a presell and an assurance that you will own the tulip as soon as they are ready at the end of the season. The futures trading got so popular, and it was referred to as wind trade since no product was involved in the initial transaction. The viceroy tulip priced at five times the cost of an average house during the 17th century.

In February 1637, the prices of tulip bulb dropped suddenly, and there was a marginal decrease in tulip trade. People were keeping on buying tulip so that they could sell at the even higher price when a society is collectively planning such a mission, and there will be a point when the supply will far exceed the demand. More people will be looking forward to selling the tulips they invested in than the number of people trying to invest in it, and this is when the bubble bursts.

This event made the people rethink how value is decided for a particular commodity, at that time it was crazy for a flower to be sold at that kind of prices. The event greatly influenced the thought process of the people by changing the perspective of how things work. This Tulipmania is in line with the Greater fool theory which states that people buy products at a foolishly high price with the expectation of selling it to a "greater fool" later.

Another notable bubble is the dot-com bubble which occurred when the internet was getting adopted, and the usage of the internet was increasing significantly. Let's familiarise with the dot bubble in a chapter exclusively dedicated to it.

The bubbles are formed when the value of a product, company or any asset deviates from its intrinsic value. Intrinsic value is the value of the product determined by fundamental analysis, and it can be seen as the actual value of the product, the value which is based on the real application of the product. We cannot predict the formation of a bubble nor can we try to burst it prematurely, any attempts will cause economic chaos. It is advised to let the bubble pop on its own, and the impact should be dealt with after the event. These bubbles can destroy a great deal of wealth which is following the debt deflation theory proposed by Irving Fisher. Another effect of the bubble is its influence on spending habits. When people are in possession of assets with high value, they will be wealthier and thus spend more to match the vibe of wealthiness. When the bubble burst and the value of the assets drop, these people will decrease their expenses because they do not feel wealthier anymore. This process will thus decrease the overall economic growth.

What causes a bubble?
We can run a post-event analysis on situations like tulip mania and deduct some possible explanations, but it is somewhat unclear what causes a bubble.

Bubbles can happen even when there is no greater fool theory involved or when there is no bounded rationality which states that the decisions are influenced by our cognitive abilities, controllability of the problem and the time specification. The bubble can even occur in unpredictable markets or when crowd speculation is not involved.

When there are more options for liquidity, there is an excellent movement of assets. When this liquidity is backed by low-interest rates there can be chances of the bubble because more people get into investments of promising products and the investment aims at a particular asset. The solution is to increase the interest rates so that the investors become aware of the risk involved with the investments.

Bubbles can also be caused because of extrapolating previous trends into present situations. The herd mentality of the crowd can also drive the formation of the bubble. The people are investing in assets based on its market movement. The herd behavior is driven by the combination of speculation(an expectation that the asset will increase in value) and reciprocation(others are finding the asset valuable therefore I will be investing in it as well).

There are several characteristics associated with a bubble like borrowing on a large scale to invest in the asset, a decrease in interest rates to encourage more investments, large-scale spreading of attempt rationalizing the event and the thing which we are easily observing is the adoption by the media which exploits the mass speculation.

Summary
- The exotic tulips were facing massive speculation that their intrinsic value was meager compared to speculative value.
- Even though we can understand the factors which caused most of the already occurred bubbles, we cannot accurately predict future bubbles.

9. Dotcom bubble

Preview:
- The crash of Pets.com and Webvan
- The fall of on-paper Millionaires
- Introduction of the internet to public
- The aftermath of the bubble burst

"The stock market is a discounter of all known information."
- Kenneth Fisher

Pets.com was a company opened in 1999 which sold accessories and supplies for pets over the internet. It ran massive marketing campaigns, and the image of pets.com was stuck in people's minds. The campaigns were so massive that It even appeared in 2000 Super Bowl ad and the mascot of pets.com was interviewed for the magazine - People, and it appeared on the show Good Morning America.

The company failed to deliver what was promised. All the attention and awareness of the company was futile when the company was losing money while selling their products. The company was spending three times the profits it made; the company lacked a proper business plan. When you do not have a good business plan, and you spend enormous amounts of money on marketing campaigns, you are building a ticking company which could implode anytime. It was not even sure whether there exists a market for the pets accessories niche. The extra discounts and the cost cuts for increasing the customer base were not feasible as the profit margin was very less. When they realized that further investments could not be obtained, it was decided to sell the company. When the liquidation of the company was announced, its stock market prices plummeted from $11 to a mere $0.19, considered as one of the greatest dot-com crash.

Another case of dot-com crash victim is Webvan which was an online grocery store shut down on 2001 after operating for three years. Upon the realization that the company had a beginners advantage, there was a massive influx of investments flowing in. In 2000, the expenses of the company were more than double its profits, and it was the time when the company was operating at its peak. When the company stopped the service in June 2001, the cumulative lose was around $800 Million. The unsold commodities were distributed to the local food storages and food banks. About 2000 people lost their jobs as a result of this shutdown.

The failure of the company was mainly attributed to the attempt to expand fast using hefty amounts of investments without any proper business plans. Webvan picks the wrong audience for their product who couldn't deliver them enough profits. Everything was built from scratch(like a warehouse), there was no minimalistic approach used anywhere which massively increased their expenses.

Internet was made easily accessible to the public with the introduction of Mosaic web browser in 1993. By 1997, the number of personal computers in the US became about three times when compared to 1990. From 1997 to 2001, the internet was getting adopted at a high rate, as it turned into a necessity as opposed to a luxury which it was earlier. Like any new technology, the internet experienced a deal of public speculation. People were pouring money into everything which had a .com tag attached to it. It was straightforward to obtain funds for your company from IPOs during this period. Many existing companies attached an extra .com to their names to get funded, and it worked!

The sudden influx of increasing demand for internet-based companies changed the outlook of investors. The investors were expecting the companies to be money-making machines if they moved in faster and exploited the beginner's advantage. Stories of people who quit their jobs to dedicate their entire time in day trading were widespread.

People were instantly turning into millionaires when the companies in which they had shares underwent IPOs. The deviation from traditional investing trends prompted the investors to invest in the company which never made any profits. The only condition for the companies was to look promising and to give hope to the investors about the riches they are going to make. The early employees received shares of the company they were working in, but most of them were not able to sell soon enough because of the lock-up period of the shares. Market Cuban leveraged the situation by successfully using hedging to protect his profits.

We know how the sensation of wealthiness can prompt people to spend more. Most of the people were millionaires on paper, that was just enough for them to spend extravagantly. The newly formed companies like Webvan were investing heavily in advertising and marketing so that they could profit from the increased brand awareness and the beginner's advantage. The massive investments and the lack of solid business plans resulting in the expenses overshooting the profits. They just wanted to build brand recognition such that they could cash out the profits later in the future. During 2000, Super bowl commercials included 16 .com companies commercials. This means they were spending about 2 Million dollars for about 30 seconds of screen time. Dotcom party was a great attraction during this period which was an expensive party conducted to mark the launch of a dot-com company's product or during a venture capital funding. The justification for throwing such parties where marketing, publicity, and recruitment but the reality was that the real effect of conducting such parties was not recorded. It was not sure how much it benefited those companies, the critics often considering this as a useless display of wealth. Most of the people who attended the parties were just in for fun and rarely cared about the real purpose of the party.

After the hype started to die out, venture funds were not available. Investors were getting more cautious with their investments as more and more dot-com companies were coming down and many were liquidated. The industries which greatly benefited from the dot-com bubble like the marketing and shipping agencies reverted to their standard scale of functioning. The lifespan of the remaining companies was measured by the burn rate of their funds. The companies which survived the dot-com crash were valued very low. When the whole mess settled down, and the internet-based companies were stabilized, many of those who survived the crash become very dominant in their respective sector. Amazon, Google, and eBay include the successful survivors of the dot-com crash, who are now on the fortune 500 lists.

Summary
- Many dot-com companies failed due to the absence of solid business plans
- In 1993 the Internet was readily available to the public through a mosaic web browser
- In 2000, Super Bowl included 16 dot com company commercials.
- After the bubble burst investors were more cautious about their investments
- The survivors like Google, Amazon, and eBay found great success in their verticals

10. Surviving the crash

Preview
- Survivors of the dot-com crash

"Your fear is 100% dependent on you for its survival."
-Steve Maraboli

What doesn't kill you makes you stronger, this is very true in the case of the dot-com bubble. The bubble did try to kill off every company. Some companies were wiped off, and some survived with a bit of luck and some struggle despite their share values dropping very low. Let's see how some of those survivors made their way through the crash.

The largest retailer Amazon was founded by Jeff Bezos in 1994. Amazon was more of an online bookselling portal when it started off. When the IPO was launched, the shares were sold at $18 which rose up to $100 later on. Just like every other dot-com companies at the time, Amazon was focused on building brand recognition and not on profits. Amazon could have been destroyed in the dot-com crash, but the decision to sell bonds to investors oversea saved the company. The deal was completed just one month before the dot-com companies came crashing. This deal doesn't mean that the deal made amazon crash-proof, the Amazon stock fell to $9 from $100. If they are waiting another month for this deal to happen, it would have been tough to survive the ride. The main reason which backed the survival of the Amazon is the viable business plan and a reasonable timeline with proof of profitability. The other companies like Webvan and pets.com mainly failed because they failed horribly in creating a good business plan for their survival. The crash also wiped out the competitors increasing their market share even if values were falling. A good plan combines with an essence of luck which played out at the right time.

eBay was popular on its auction trend, and the company was founded by Pierre Omidyar in 1995. By the beginning of 1997, the number of auctions conducted on the website reached up to 2 million. You could see anything on eBay, and it was a very diverse marketplace selling the even real estate. The option to auction the products was well welcomed by the people who helped them survive the dot-com crash.

Netflix always served the need of people by providing an affordable source of watching TV shows and movies. It was something for which people always had a craving. The business plan architectured around satisfying the craving of the people did help them to survive the crash. Even when Netflix was finding success in DVD rental, their future vision kept them investing in streaming the shows online as the quality of the internet was sure to increase in future.

We spend a lot for travelling, and when we are on a limited budget, we try to cut small the total expenses by finding discounts and negotiating wherever we can. Priceline.com made these travel-related discounting processes easier by letting people find suitable discounts and call out for their preferred price. The stock price of priceline.com reached up to $86 which fell to less than $10 during the crash. The company thrived with its name your price strategy and its significant acquisitions which helped it expand the market.

Internet was a relatively new thing even at the time of the crash, everyone who hoped in the hype train with hope, ambition and dreams met with colossal failures as their business plans didn't give them a solid ground to stand on. Those who became paper millionaires came down as fast as they raised high. The dot-com crash can be seen as a real test to filter out the best among us.

Summary
- Even after the stock price of Amazon fell to $9 from $100, it survived the crash because of the decision to sell bonds overseas and its solid business plan

11. Time Machine

Now I'm going to give you a time machine with a one-way ticket to 1990. You are now going with all the knowledge of how internet became what it is today. You have the precious knowledge of how Google, Amazon, and Facebook are a big deal now. So with all this information, there are many ways you can get it capitalized.

Build those on your own and invest heavily in them during early stages!

The building is tricky even with all the knowledge you have because you only know everything from what media is telling you about those giants, you don't have a clear picture of what they had to go through and how much resourceful they had to be for those giants to be what they are now. Well if you trust your knowledge enough to build competition and if you are sure about it being successful well sure go for it.

The easier option is to become an early investor in these companies. You find these people somehow and make them an excellent offer which they cannot refuse and voila you have yourself a sweet deal of a future multi-billion dollar company.

So, am I giving a time machine? ...Yes, I am.

It is not for going back to 1990 though; I'm giving you the opportunity to see what will shape our future. It is like the future you build a time machine which was risky to use, so you send me to convince you for investing in these technologies. By investing, I do not mean put all your resources in every place you can find the name of the blockchain, AI and IoT. That was the purpose of introducing you to the concept of economic bubbles and what happened after the internet crash. You can either buckle up and start conceptualizing your idea and build something from the start. For that ride, you need to be ready and resourceful to survive all the crash and burns to cash out the future fortunes. The other option is to find something promising in these fields, which can be a promising startup with a convincing and promising plan with a great team to make it happen.

You already have the advantages of the time machine, right when you bought this book. Now, all you have to do is the act. Read on and enjoy the journey!

INTRODUCTION TO THE TRIO

1. Introduction to AI

AI must have caught your interest beforehand, even before you picked up this book, thanks to all the movies and comics which gave a cool image to artificial intelligence. Did you desire to own a Jarvis of your own as shown in Ironman? Did Wall- E make you wish to own a pet robot? Or maybe you are the kind of person interested in resurrecting the Frankenstein monster, in some way or the other, AI did not fail to catch your attention. If AI already has your attention, we are already one step closer, let us get into a bit of exposure to AI, before we start let's look at the upcoming chapters and why you should read it.

The basics of Artificial Intelligence originates from our thinking capability, get a better base on artificial intelligence by understanding human cognition in "What is AI"

Why don't we hear a great deal about the progress in the AI field as much as we hear about the expectations in the AI sector? Is it because there is very low progress? Is everything hyped up and the progress hardly meets expectations? You can find the answers to that in "AI Effect"

When the idea of imparting the cognition of humans to machines started to bud, that is when the idea of artificial intelligence originated. Learn all about the timeline of artificial intelligence in "History of AI"

Everything about how the intelligence of these machines can be compared to human intelligence can be seen in the chapter "Philosophy of AI". Can the machines turn against us? How far can we go with progress in Artificial Intelligence before it starts backfiring on us? Find the answers in the "Ethics of AI".

Have you experienced a weird feeling when you come across an artificial thing like a doll or an animation which resembles a natural thing? Why does that happen? How can that be avoided so that we feel more comfortable among physical AI machines? Find all those answers in the chapter "Uncanny valley".

The old question of "Will AI steal my job?" is addressed in the chapter of the same name and I can ensure you that you will start feeling optimistic after going through that.

When machines were designed in the beginning, the intention was to make them perform things which are already taught to them. When we understand we can make the machines learn, it opened up a wide door of opportunities. Learn more in the chapter "Machine Learning".

The replication of human neural systems can be seen evidently under "Deep learning" which comes under machine learning. This is a great example of how nature brings out the best of our innovations.

So what's going in the AI scene now, get a better picture by reading through the "Major AI initiatives of big companies"

Now you have the structure of upcoming chapters set in your minds, so let's start exploring the Artificial Intelligence.

2. Introduction to blockchain

The blockchain is an immutable record of data which is shared throughout the network which is related to that particular data. What can an immutable record of things provide? We know how data is getting manipulated for covering up cash siphoning from different industries. Data manipulated decreases the validity of audits. If there is a system which offers to keep a record of data is immutable, it will give a better transparency in the system where manipulations can be uncovered easily and the process of audit ting becomes way more meaningful. Read on to understand what chapters are we dealing with in this section and why should you read it.

The bottom line for implementing blockchain is because of the trust issues we can face in the current system. We need trust among the participants for the network to function properly. This idea can be understood by the Byzantine generals' model, which generalizes issue faced by a network with multiple members.

Satoshi Nakamoto introduced the concept of blockchain through his whitepaper. Who is Satoshi Nakamoto? No one really knows the real identity of Nakamoto, to more refer to the chapter dedicated to Satoshi Nakamoto.

How can you process a bitcoin transaction? Does it work like a normal cash transaction or should you be doing some extra work to make the transaction happen? A bitcoin transaction is nothing like the normal cash transaction, refer to the Bitcoin transaction chapter to learn how you can enable a bitcoin transaction. After you understand how you made that transaction work, now you must be wondering what was going behind the scenes of the transaction and what is really happening when you process the transactions? The chapter "How Bitcoins works" answered these questions for you.

The wide spectrum of blockchain which could revolutionize normal working of the world was introduced by Ethereum. If you are too familiar with bitcoin and you did not give any chance to other promising cryptocurrencies, you can start by "Intro to Ethereum, which comes next to bitcoin in market capitalization. You can also use this chapter to brush up Ethereum knowledge.

Read the chapter dedicated to Smart contracts to understand how Ethereum enabled the wide range of blockchain applications. You can even start contributing to the blockchain sector by starting to know more about Smart contracts which are the essence of the blockchain.

You must have heard about Turing completeness introduced by Alan Turing, how it applies to Ethereum and why it is important to implement Turing completeness to Ethereum network. Find the answers in "The Turing complete Ethereum network"

The concept of DAO looked very promising but in the end, it did not quite deliver what was promised. What happened with DAO? Read more on "The hack on DAO".

Now that you have learned about what blockchain can do let's go into the details of "How blockchain works". This is more of a brushing up basics of what you caught in previous chapters. Let's go into the details of features of this promising tech in the chapter "Features of Blockchain".

The application of blockchain is enormous with both private and public blockchains available to play around. What are those? And what should you choose for your particular situation? Learn more in the chapter public vs Private blockchain.

Everything has its downsides and dim areas, explore the "Limitations of blockchain" and be aware of the limitations that come along and the possible solutions which are under development for solving these issues.

Now that you got a flavor of what is coming, let's dive into The blockchain.

3. Intro to IOT

The best part about the internet of things is that they can remove the human intervention in many intermediate steps and make many processes far more efficient. Using IOT you can pre-program your oven to preheat your food just before you reach home after office. You can use location data to feed the system which will calculate the time of arrival using the dynamic road traffic data and voila food is hot and ready to be served by the time you reach home. So now let's see the objectives of each chapter in this section.

Get more familiar with the concept of IOT by understanding about its origin and a peek into how IOT can make a change and create an impact in the "Introduction".

There are different types of IoT depending on the target market it addresses. The chapter "Types of IoT" describes and differentiates the different types of IoT.

We know how the concept of the internet of things works by now but what are the components of the system which runs IOT? How is the data collected from different environments and how is it communicated across different machines? All this are to be unveiled in "Elements of IOT".

IOT sure does make a lot of stuff easier but what is the downside? What are the risks associated with everything being interwoven? What happens if there is a crack in the system? Read "Security" to find the answers.

In "Future of IOT" find out Where do we see IOT in the future? What are the initiatives taken by different companies to make better IOT integration?

Well, that's all for the summary of IOT chapters, good luck with your IOT expedition.

AI GUIDE

1. What is AI?

Preview:
- Defining Intelligence
- The human cognition

"Artificial intelligence would be the ultimate version of Google. The ultimate search engine that would understand everything on the web. It would understand exactly what you wanted, and it would give you the right thing. We're nowhere near doing that now. However, we can get incrementally closer to that, and that is basically what we work on."
—Larry Page

What is intelligence?

Intelligence can be explained merely as the ability to use a wealth of information which is learned earlier to apply it in a future situation. It can be seen as the capacity to use the available resources. Intelligence itself encompasses a broad spectrum of factors from being self-aware to solving a problem.

Imagine you are facing a problem, you are on a weekend forest expedition on your own. Suddenly you hear a growl, and you find a gigantic hungry bear ready to pounce on you. Let's see what all factors increase the chances of your survival.
Understanding the situation, the magnitude of the threat involved.
Self-awareness to understand the experiential knowledge you have at your disposal.
An emotional command to not let your fear and anxiety take over clouding your mind.
Awareness of the possibility and reach of resources to contain the situation
Using logic and creativity to fabricate a plan.
The efficiency of solving the problem practically.

In this situation, the intelligence can be measured as how efficiently you can escape this bear situation. Every problem we face on a daily basis and the ones which we come across on a rare basis requires the combination of most of these factors which helped you contain the bear situation.

Intelligence isn't just a problem-solving tool, and it can also be used to make a situation better.

We, humans, pride ourselves on the level of our cognitive abilities which created unity among us like no other species. Suppose you are in a new town and you happen to talk with a stranger. When you learn that he belongs to the same hometown as yours and shares the same culture, it invokes a sense of familiarity. This feeling of familiarity which we feel with strangers because of the collective myths we believe in is what helps us form bigger networks as compared to other animals. The cognitive functions also include the high levels of motivation in humans. Why do you think we feel motivated? Isn't it derived from the very fact that we feel connected to others somehow?

If we humans are capable of achieving this level of cognition, can we replicate it to something we are creating? This very question is the basis of Artificial intelligence. Using Artificial Intelligence, we are expecting the machines to portray the cognitive functions displayed by the humans. This concept was well taught in movies like "Terminator" and "I robot" which shows machines which were so advanced that it became self-aware. In "Terminator," the height of self-awareness was tremendous that the machines decide to eliminate the humans. These stories do sound crazy and frightening. How much of this is possible? Can machines become self-aware? If they can become self-aware, in how much time will that happen? Can machines wipe out the human race in the future?

A floodgate is now opened, and many questions are rushing in, but how can you find answers or at least how you can increase your chances of finding the answers. The answer, we have already talked about which is Awareness. How can you increase your awareness? By exposing yourself to information. To give you a head start let's learn about Artificial intelligence in the following chapters and increase our awareness.

Summary
- Intelligence is the ability to use previously learned information in a future situation.
- Human cognition can be used as a baseline to develop artificial intelligence.

2. AI Effect

Preview:
- Why is the AI effect like a magician's secret?
- Why do we see more projections in AI than signs of progress?

"Some people call this artificial intelligence, but the reality is this technology will enhance us. So instead of artificial intelligence, I think we'll augment our intelligence."
—Ginni Rometty

What do you feel about the things you don't know?

What do you feel about things you haven't experienced?

What are the feelings we can associate with something very unfamiliar?

Suppose you are stranded on an island. You are out of food supply which hardly lasted a couple of days. Within this span of time, you explored every bit of the island except the dark cave which is in front of you now. You "don't know" what is inside the cave, maybe a ferocious animal or the cave is filled with suffocating poisonous gas. The cave is very unfamiliar, and you feel FEAR.
Next, let's consider a magic show, the illusionist displays a bird and shows nothing is hiding up his sleeve. Now he makes the bird disappear by compressing the cage. As we are in awe, wondering where the bird disappeared, he pulls out the bird from thin air. We "Don't know" what is happening here, but are we feeling scared like the previous situation? No, there is a sense of amusement, interest, and curiosity invoked.

The AI effect is very similar to a magician's secret. The bird illusion which was mentioned above is seen in the Christopher Nolan's movie "Prestige." In the movie after the illusion is performed, an intelligent little boy in the crowd understands the act and starts crying. He points at the reappeared bird and tells; the magician killed the bird,s brother. The trick involved the killing of one bird, to make it vanish. Later the illusionist explains to the kid that the value of the magician is in concealing the trick. Once the magician gives away his trick, the people lose the value for magician which they had before he revealed the trick.

The same thing happens in Artificial intelligence. Various discoveries are bringing us one step closer to the future of artificial intelligence, but this development suffers a great paradox. When something new is discovered related to AI, it stops being magical just like the magicians trick. We claim that it is not real intelligence but rather just computation, as stated by Rodney Brooks.

We consider the new solved problem not to be a part of AI. If we examine the AI field, we can see that everything associated with AI is something we are trying to solve. The discoveries which were done in this field are integrated into other fields and applications without giving much regard that it is an AI-based solution. These solutions are seen as if they are normal and a part of this field which adopted it. In short, we can say other fields are stealing away the achievements of the AI sector.

This phenomenon due to the quick commercialization of a new product as soon as artificial intelligence solves a problem or improves an existing solution. The newly formed solution will be identified with the product than the artificial intelligence sector and thus leaving us with the question of why AI is not showing any drastic advances. We often fail to realize the only thing which motivated us to ask such a question is the methodology of tagging. For instance, the implementation of voice recognition in mobile phones have improved over time, and today it is seen as an integral part of every smartphone which creates a perception that it is the development of smartphone industry rather than AI.

There is a hype around AI technology which can't be denied. This hype is often misused for acquiring fundings for startups, businesses, and projects. This exploitation of AI was very evident after AI started recovering from its lowest point in 1990.

We, humans, hold ourselves in very high regard. We consider ourselves very advanced relative to anything else on the planets. If we are thinking on a very narrow perspective, every individual wants to individual and should possess some great quality which can distinguish them from others. Now imagine, an enormous group of these individuals(the human beings) powered by superior cognitive abilities(like we claim). This group would want itself to be considered unique and in very high regard. This collective complex of the human species might be the reason why we deny the possibility of something can be as good as us or is on a path to become as superior as us.

Summary
- When there is an innovation with AI, it is quickly associated with other sectors that the credit is mostly not given to AI itself.

3. History of AI

Preview
- From Frankenstein to Google Duplex

"Artificial intelligence is growing up fast, as are robots whose facial expressions can elicit empathy and make your mirror neurons quiver."
—Diane Ackerman

The concept of Artificial Intelligence is so ancient that it is mentioned even in mythologies. Talos was a considerable automaton(A machine which operates without any external aid) formed from bronze. In Greek mythology, the purpose of Talos was to protect Europa, the mother of King Minos of Crete. Other examples of artificial intelligence in Greek mythology include Galatea, a statue made out of ivory which came to life and Pandora who was made from earth, first woman created by the Greek gods.

Many cult images are used around the world, which is said to possess a conscience and is taken care of regularly. These cult images are the most important figures in a temple. Even Though this sounds mythical, it is for sure based on the concept of artificial intelligence.

Mary Shelley's fiction Frankenstein can be seen as the debate on morality while building a highly cognitive artificial intelligent being. We can learn several things from Shelley's novel like:
The experiment should not be conducted in secrecy and isolation so that there is more control over the entity which is being created
The creation should be observed, trained and primed so that it won't be running amok as the Frankenstein monster did
The community should be prepared for acceptance of such a creation(You don't want the community to oppose your experiments because of uncanny valley experience)

The word robot was first used in Rossum's Universal Robots, a play by Karel Capek. Game theory was very crucial in the development of AI. Game theory was introduced in 1944 by John von Neumann and Oskar Morgenstern.

In 1950, Alan Turing came up with the Turing test, which measures how different is an artificial interaction compared to that of human interaction.

In 1950, Isaac Asimov published a set of 3 rules, which had a significant impact on artificial intelligence ethics. The three rules are:

A robot may not injure a human being or, through inaction, allow a human being to come to harm.

A robot must obey the orders given it by human beings except where such orders would conflict with the First Law.

A robot must protect its existence as long as such protection does not conflict with the First or Second Laws.

The significance of this law is that even if the artificial intelligence reaches the level of humans, the safety of our species will be ensured.

In 1958, John McCarthy of MIT designed the programming language Lisp which was adopted for programming AI. In 1959 Marvin Minsky along with the same John McCarthy founded MIT AI Lab.

In 1960, the Bayesian model for Inductive inference and prediction was introduced by Ray Solomonoff which were the first stepping stones of Mathematical theories of Artificial intelligence.

In 1963, through his work ANALOGY, Thomas Evans showed that Artificial Intelligence could solve the same analogy problems presented in a traditional IQ test. Charles Vossier and Leonard Uhr released the first machine learning program, which solves many problem statements which existed at the time due to lack of adaptability.

In 1964, Danny Bobrow showed that AI could understand natural language so that it could solve algebraic problems.

In 1979, CHI system was demonstrated by Cordell Green, Davide Barstow and Elaine Kant which enabled automatic programming. The first autonomous vehicle, Stanford Cart was built by Hans Moravec.

In 1986, robot cars were made which could reach up to 55 miles per hour.

In 1997, the world chess champion Garry Kasparov was defeated by the chess program Deep Blue. RoboCup football was conducted for the first time featuring 40 teams of robots.

In 2000, The tough to reach spots in Antarctica is explored by the bot, Nomad.

In 2002, Roomba released by iRobot navigated and avoided obstacles while cleaning and vacuuming the floor.

In 2004, DARPA Grand Challenge was released, which rewards the best autonomous vehicles presented at the event. NASA explores the surface of Mars using autonomous rovers Spirit and Opportunity.

In 2005, a humanoid robot which can walk as fast as humans were released. This was ASIMO from Honda; this robot was able to deliver trays of food in restaurants.

In the same year, The project known as Blue Brain is initiated which aimed at simulating brain at a molecular level of detail.

In 2007, autonomous cars were released by DARPA which could obey traffic rules and could function in an urban setting

In 2009, Google created its self-driving car.

In 2010, the Computer Vision group at Cambridge Microsoft Research facility created Kinect which could track human body movement. Using this tech, the players could play Xbox wirelessly. This upgrade was achieved by using a 3D cam and infrared detection.

In 2013, Carnegie Mellon University released the Never Ending Image Learner which could run analysis and comparisons on two different images to find their relationship

Before 2014, the personal assistant applications like Siri, Google Now and Cortana are released which could perform custom activities like answering to user's questions, providing suggestions and performing actions for the user.

In 2017, Asilomar Conference on Beneficial AI was conducted so that the development in AI sector won't be compromising the security or bring forth any risk.

In 2018, Alibaba's AI scored higher than humans in reading and comprehension test conducted at Stanford University. Google Duplex was announced which is an assistant Ai to book appointments. According to LA Times, the voice was indistinguishable from the human voice.

Summary
- The History teaches us the change and how the past evolved into the present, the study can give us an insight on how the future can unravel from this present.

4. Philosophy of AI - the Chinese room

Preview
- The three questions which form the basis of AI philosophy
- The Chinese Room experiment and Symbol learning

"The development of full artificial intelligence could spell the end of the human race....It would take off on its own, and re-design itself at an ever-increasing rate. Humans, who are limited by slow biological evolution, couldn't compete, and would be superseded."
-Stephen Hawking

There are three main questions which are the basis of AI philosophy.

How similar is human intelligence to artificial intelligence?

To what extent can machines solve problems that human can?

Can machines function by how they feel about things?

Alan Turing proposed a method which can compare machine intelligence to that of humans; this is the Turing test. If a machine can answer all the questions fed to it as humans do, then the machine is as intelligent as the humans. We can see this in action when a random chat box pops up in some websites; sometimes they are so well programmed it is hard to distinguish a bot or a human handle that. One of the criticisms that this test faces is regarding the goal of creating artificial intelligence if the robots are meant to be more intelligent than humans, why are we trying to make them resemble human activity.

Intelligent agents are entities which use sensors to observe an environment, the performance of these agents are measured regarding how well these agents can use this observed data. These observations are also recorded and repurposed for later use, the more efficient they use this recorded and observed data that means they are exhibiting better performance. The performance is the measure of the success of the machine. This is not a definition which tries to bring the thinking capacity to resemble humans; rather it is based on handling the situation more efficiently.

John Searle's Chinese room argument suggests that however hard we try, we cannot impart the concept of mind into machines. According to the Chinese room experiment, if you are locked in a room with a giant book which has many instructions. The book contains all the replies to every question that can be asked in Chinese. You receive pieces of notes through the door gap, following the instruction of book, you will keep providing replies. The person on the other end will be convinced that you understand Chinese, this is like a Turing test for Chinese. You cannot be differentiated from an original Chinese speaker which is essentially the difference between a Strong AI and a weak AI. According to Searle, the machines act in the manner, and they cannot act like humans by being conscious of the situation. There is a clearcut difference between possessing a mind and simulating one.

The Chinese room argument doesn't mean that machines with superior intelligence can't be created, but it implies that machines which think like humans and function as human brains do can't be created. This is more of a philosophical approach to the idea of Strong AI, the researchers do not care most of the time about the categorization, but they look out for superior intelligent behavior and other results.

The Chinese room is an excellent example of how a symbol system works. There is a set of predefined instructions according to which we are acting and responding without actually understanding the meaning of those. This works in the same way that the computer syntaxes and algorithms do. The man in the locked up room doesn't understand Chinese, but the system which comprises of the man and the conversion booklet understands Chinese. Here, what matters is how the system functions.

Summary
- A machine can pass a Turing test if it can respond as humans do.
- Chinese room demonstrates the process of how machines learn through symbolic learning.

5. Objectives and Approaches

These are the different objectives of Artificial Intelligence.

Knowledge reasoning

When the step by step approach to solving problems crumbled under complex situations, the solution was to impart the basic common sense to the machines so that they can identify objects and relate it to different situations, states and the effects it can create.

Planning

The machines should be capable of approaching solutions in a well-planned manner, and they should also be aware of how their actions can potentially change the current state of things. The other factors of the system which can influence the solution to the problem should also be considered and be accounted for.

Machine Learning

This is the vital part of Artificial Intelligence where the system is improved by imbibing the machines with a power to learn from experiences. We will go into details in the chapter dedicated to machine learning.

NLP

The natural language processing is the ability of the machines to recognize and replicate the natural language of humans. This can bridge the gap between the machines and the humans to a great extent. Make sure you go through the dedicated chapter for better understanding.

Computer vision

The ability to receive inputs from corresponding sensors such that the human vision can be mimicked and can be used for judging situations based on visual information in the concerned system. For more information go through the chapter dedicated to "Computer vision."

Robotics
The robots are the machines which are replacing humans with their ability to perform human-like actions, especially in risky environments.

How can AI problems be approached?

Symbolic: This concept can be best understood by referring to the Chinese room experiment mentioned in the chapter "Philosophy of AI." The typical approach boils down all the human interactions to a symbol manipulation system. The typical approach can be based on logic or knowledge.

Sub-symbolic: When the representation is not specific enough, then it is known as sub-symbolic which can be evident from its name. This was motivated by the fact that the typical approach couldn't mimic human behavior.

Statistical: The Statistical revolution of the 1990s influenced artificial intelligence to include statistics to improve the results. The neural network is based on the statistical approach.

With the problem statements and a bit of clue here and there in our hands, stay tuned for more AI exposure.

6. Machine learning

Preview:
- Introduction to machine learning
- Deep learning, Artificial neural network, Convolutional neural network, Recurrent neural network

Alan Turing gave a new perspective of AI by making us think "Can machines do what we can do?" instead of "Can machines think?"

Machine learning is the branch of artificial intelligence were algorithms learn from the data fed into the systems. Opposed to the method of being programmed, the machines will learn progressively improving their performance in accomplishing a task. Machine learning has been a buzzword for quite some time and can disrupt many industries. Arthur Samuel introduced the term machine learning in 1959. The roots of machine learning lie in computational learning theory and pattern recognition. Instead of programming everything on a computer code based on some protocol, machine learning works by learning from large datasets similar to how human brain process information so that they can provide predictions when fed with data. In the industries, the machine learning is popularly known as the "predictive analysis" which help them to optimize the working of the industry.

Tom M. Mitchell explains the machine learning algorithm as follows:
"A computer program is said to learn from experience E concerning some class of tasks T and performance measure P if its performance at tasks in T, as measured by P, improves with experience E."

Depending on whether the system is primed, the machine learning can be categorized into supervised learning or unsupervised learning. In supervised learning, examples are provided to initiate the learning mechanism. Depending on the nature of these examples (training data), supervised learning can be further classified into Semi-supervised learning, Active learning and reinforced learning. When the system is only provided with incomplete examples, with most of the outcome possibilities missing, then the system is semi-supervised learning. Depending on the available budget, the system is fed with examples based on which the system judges other situation, and this is known as active learning. Reinforced learning happens in cases like driving a car where the information is given accordingly to the actions of the users; this process is dynamic. In case of unsupervised learning, we don't train the system, but instead, the system has to decipher the data on its own

The objective of the machine learning is to generalize situations such that it can predict future situations. The algorithms are developed based on the computational learning theory which falls under theoretical computer science. The training period of the machines is mostly predefined which means that the results can be only be expressed in a probabilistic model.

Deep learning is the subdivision of machine learning which is inspired by the human nervous systems. The deep learning models are designed based on the communication and information processing patterns of the human brain.

Deep Learning:

The machines can use deep learning for categorizing and classifying data by themselves. Using deep learning the machines can identify images and videos with dogs without explicitly programming them to identify dogs. The deep learning uses a humongous amount of data so that it can learn from the context of the image so that it could be repurposed for future prediction.

Before going to deep learning, you should be aware with the term artificial neural networks. Artificial Neural Networks as the name suggests, it is inspired by biological neural networks inspired by the human brain. The brain is a natural supercomputer that makes humans intelligent and enables them to take actions and decisions accordingly in different situations. What if we can make a machine which can assess situations and take decisions accordingly as the brain does? Will it be super cool? Artificial Neural Networks are designed in this objective; they are used to attain cognitive capabilities in machines by replicating the structure and function of the human brain. If these networks are complex and more profound and we use them to learn complicated things, then it is called deep learning.

Artificial Neural Network:

Thus, now we know that artificial networks are designed to replicate the functioning of the human brain, let's understand how does it achieve this? An Artificial Network has two major components, the nodes, and the edges. Similar to neurons in the brain, Artificial Neural Networks have a collection of connected nodes called artificial neurons which is the base of the network. The connection between each of these nodes is called "edges." Further, the signals are transferred from one node to another through edges.

Generally, in Artificial Neural Network implementations, the signal transferred between artificial neurons is a real number, and the output of each artificial neuron is computed by some non-linear function of the sum of its inputs. Artificial neurons and edges typically have a weight that adjusts as learning proceeds. The weight increases or decreases the strength of the signal at a connection. Artificial neurons may have a threshold such that the signal is only sent if the aggregate signal crosses that threshold. Typically, artificial neurons are aggregated into layers. Different layers may perform different kinds of transformations on their inputs. Signals travel from the first layer (the input layer) to the last layer (the output layer), possibly after traversing the layers multiple times.

The deep learning can be further classified into Convolutional Neural network and recurrent neural network.

Convolutional Neural Network

Have you ever thought how amazing the sense of vision is? Humans can identify objects within a particular field, in a fraction of seconds. Apart from identifying the object, we can also sense their depth, correctly distinguish their contours, and separate the objects from their backgrounds. Somehow our eyes take in raw voxels of color data, but our brain transforms that information into more meaningful primitives—lines, curves, and shapes—that can give further details about the object.

Convolutional Neural Networks, inspired by the human vision is a subset of deep learning which is used in analyzing visual imagery. CNN makes use of feed-forward artificial neural networks which means that the information flows in only one direction from the input nodes to the output nodes through the hidden layers. They are also known as shift invariant or space invariant artificial neural networks (SIANN), based on their shared-weights architecture and translation invariance characteristics.

They have got a wide array of applications in image processing, video processing, recommender systems, etc.

CNN's are used to process data having a grid-like topology form and uses a linear mathematical function called convolution in between the network layers. A CNN consist of input, output and multiple hidden layers. The hidden layer constitutes the convolutional layer, Rectified Linear Units layer, pooling layer, and Fully-connected layer.

Recurrent Neural Network

Another popular deep neural network is the recurrent neural network. When compared with feedforward networks RNN can process sequences of variables using their internal state or memory. They are widely used in applications such as language modeling, handwriting recognition, speech recognition and so on. The layers in RNN is intended to bring in memory and not hierarchical processing.

According to Occam's razor, while solving a problem, the simplest solution is the right one. Even though they are better processing evolving in machine learning to get better predictions, the simplest of them will also prevail due to this theory. This is used while developing heuristic models of machine learning, even if the complicated models can bring in about a 1% increase in result accuracy.

Summary:
- Machine learning can be supervised learning or unsupervised learning
- The objective of machine learning is to predict future outcomes.
- The human brain inspires artificial Neural Networks
- Deep learning can be divided into convolutional neural network and recurrent neural network.
- The convolutional neural network is inspired by human vision
- Recurrent neural networks are used in speech and handwriting recognition as well as language modeling.

7. Deep learning networks

"Artificial intelligence is the future, not only for Russian but all of humankind. It comes with enormous opportunities, but also threats that are difficult to predict. Whoever becomes the leader in this sphere will become the ruler of the world."
                    -Vladimir Putin

How are deep learning frameworks designed, trained and validated?
The answer is the Deep Learning Framework. These frameworks provide an interface for building deep learning networks using high-level programming resources. When deep learning is applied in sectors like NPL, computer vision, social network filtering, audio recognition, and board game programs, results produced were better or comparable to human counterparts. Let me introduce you to some of the great deep learning frameworks.

Tensorflow

Tensorflow is an open source python library developed by Google for dataflow programming. It is a symbolic math library and was created for tasks which require heavy numerical computation. Since machine learning involves repeated computation of complex mathematical expressions, TensorFlow is used in applications involving machine learning and neural networks.
Thus, Tensorflow is used by developers to design and train deep learning algorithms. Tensor flow is faster than pure python code because of c/c++ backend and uses a data flow graph structure for performing the computations.

Why use Tensorflow? There are many libraries available for developers apart from Tensorflow. Some of them are Theano, Torch, Keras, and Caffe. But what makes Tensorflow unique when compared to others? Let's have a look at some of the reasons
Tensorflow has an active developer community. The framework is written in Python which has a more comprehensive adoption among the developer community. Tensorflow has built-in support for deep learning and neural networks and has a collection of Mathematical functions that are useful for neural networks.Further, Tensorflow is versatile; it is compatible with many variants of machine learning.

TensorFlow provides both python and C++ API (Application Program Interface), but Python API is more complete and is easier to use. Further Tensorflow has greater compilation times compared to other deep learning libraries. TensorFlow supports CPUs, GPUs and even distributed processing in a cluster.

Tensorflow structure and the Data Flow Graph: Tensorflow structure is based on the execution of a data flow graph. A data flow graph has 2 units: the node and the edge.

Node: The node represents a mathematical operation. It can be any mathematical operation like multiplication, division, addition, etc.

Edge: Edge represents data that is communicated from one node to another. Data in Tensorflow is represented as a multi-dimensional array called tensors.

Tensorflow architecture: Flexible architecture allows performing computation on one or more CPU, GPU, desktop, server or even in a mobile device. Tensorflow has got a tailor-made ASIC called TPU. Tensor Processing Unit or TPU is an AI Accelerator ASIC specially designed for neural network programming using the Google Tensorflow framework. They are used in machine learning computations which does not require extreme precision. Since it does not require extreme precision, it uses Instruction Set Architecture as low as 8 bit. At the same time, the simplicity of 8-bit calculations helps them to deliver higher IOPS.

Google uses TPUs in various applications like Google Photos, RankBrain, Google Translate, and Google Street View. A single TPU can process about 1 million photos in Google Photos. TPUs were not commercially available until in February 2018 Google announced they would make it available for other through their cloud computing service.

TPUs are also suitable for bitcoin mining since it can perform a large number of computations with lesser power consumption than GPUs and ASICs. Further, in the same neural network, they are 10 times faster than a GPU and 6 times faster than a CPU.

Theano

Theano is another python library used to define, optimize and evaluate mathematical expressions using packages like NUMPY and SCIPY. Theano provides an essential set of functions for building deep neural networks. The machine learning group created Theano at the University of Montreal.

Theano lets you define and evaluate mathematical expressions with vectors and matrices which are rectangular arrays of numbers. If you use Theano, you will have to build a deep net from the ground up. The library does not provide complete functionality for creating a specific type of low net. Instead, you'll need to code every aspect of a net like the model, the layers, the activation, the training method and any particular method to prevent overfitting.

Keras

Keras is a high-level neural networks library, written in Python and is capable of running on top of Tensorflow, Theano, Microsoft Cognitive Toolkit or MXNet. Keras is user-friendly and straightforward to use. It focuses on being modular, and extensible, and was designed for people to conduct proof of concepts and active experimentation with Deep Neural Networks.

The primary author of Keras is a Google engineer named François Chollet. It was developed as a part of the research effort of project ONEIROS (Open-ended Neuro-Electronic Intelligent Robot Operating System). Further, Keras is gaining extensive support from Google.

The design principle of Keras has the following characteristics:

Modularity: Model is a sequence of standalone modules that can be plugged together without any restrictions. For instance - Neural layers, cost functions, initialization schemes, optimizers, and activation functions are all standalone modules which can be plugged together to create new models.

Minimalism: Each module in Keras has to be short and straightforward.

Extensibility: It is simple to add new modules in Keras.

The core data structure of Keras is a model, which is a way to organize layers. There are two types of model,

Sequential model: This is just a linear stack of layers used to implement simple models

Functional APIs: These are used for more complex architectures, such as models with multiple outputs and directed acyclic graphs

From all the deep frameworks we went through, we can understand that each frame is offering something unique. Deep learning can be experienced in many applications now, how do you think Amazon is recommending products to the customers? The impact of deep learning is so significant that in Amazon about one-third of the purchases come from the suggestions generated through deep learning. Youtube uses deep learning for suppressing the contents which are not in line with youtube guidelines; this mechanism has indeed made youtube friendly for everyone. Uber also uses deep learning mechanisms to provide you the best and efficient means of transportation. Even Netflix movie suggestions are based on deep learning. Everything which delivers you tailor-made suggestion is mostly based on deep learning!

8. Natural language Processing

Preview
- Imitating the language of humans
- Origin of Natural Language Processing

"The voice that navigated was that of a machine, and yet you could tell that the machine was a woman, which hurt my mind a little. How can machines have genders? The machine also had an American accent. How can machines have nationalities? This can't be a good idea, making machines talk like real people, can it? Giving machines humanoid identities?"
— Matthew Quick, The Good Luck of Right Now

Natural languages are developed naturally in human beings through constant communication throughout the ages; they can come in the form of speech or sign language. Their formation is not planned like the ones we use for programming computers.

If the machines are expected to interact with humans, they should be capable of understanding human language and the data created in human language. Without the capability of recognizing the speech of humans, understanding and generating the natural language, the level of interactions will be considerably low.

The earliest form of NLP can be traced back to 1950 when Alan Turing proposed the Turing test. Turing test can be used to distinguish humans from machines by observing their responses to a particular set of questions. By the end of the 1980s, statistics had on influence on NLP, and it started relying on machine learning for yielding better results.

There are many advantages of implementing machine learning in NLP. The parameters need not be focused down when is system is capable of learning; the system will self-adapt such that different parameters are provided necessary weight. There is always a limitation to the accuracy when the rules of the system are predefined; the results can only be made more accurate by changing the basic algorithm of the system. With machine learning, the system can become more and more accurate when it is fed with more examples.

There are many ambiguous situations in languages which need to be correctly identified and tagged accordingly. This is very complex in the case of Chinese because it is a tonal nature when used as a verb. Another complexity is the ambiguous nature of punctuations which can mean different in different situations. Coarticulation is a feature of natural language which eliminates the spacing between words and thus blending the words. It is a great challenge to convert these analog signals which are blended up into discrete characters.

Facebook has developed a tool known as Deeptext which identifies the words we use based on its context to decipher its meaning. The neural networks can understand how the meaning of the words change when the surrounding text changes. This tool is used by Facebook to drive contents which can sell people products by profiling people with textual analysis.

Summary:
- The natural language processing was developed as a result of repetitive communication among humans
- The machines should adapt to natural language to increase the interaction level with humans.

9. Computer Vision

Preview:
- Can machines see?

"We can only see a short distance ahead, but we can see plenty there that needs to be done."
— Alan Turing

We, humans, can make decisions based on the things we see when this capability is replicated in the machines enabling them to make decisions. Digital images or videos are to be acquired, analyzed and understood so that the machines can make decisions like our vision works and this is what computer vision is all about.

The decision produced by the machines maybe symbolic or numerical which are derived from the profile of captured images. The profile of an image is built using its geometry, laws of physics, statistics and improved using machine learning.

In movies like Terminator, we have seen the perspective of the terminator where it identifies objects and shows attributes of those objects by the mission programmed to the terminator. That can be seen as the application of computer vision, but it's not just that we can restore images with this tech. With the learning capacity of the machines, they can fill out the distorted parts of the images when the original image is provided. Different objects can be identified, and their motion can be tracked, open the camera in your phone, and you can see the face recognition in action. This can also be used for 3D profiling of the objects, which can be used in advanced video editing.

So in brief, we are trying to automatic what human vision can do by using computer vision. Any who can be considered as an image can be included in this sector irrespective of whether it comes from different sources like multiple cameras or different sources like non-invasive medical scans.

The origin of computer vision can be dated back to 1960s, one of the budding stage projects was a camera attached to the PC which can acquire images, and the computer had to describe what the captured image was.

How do we see objects? The light reflecting from the objects are responsible for it, so for the machines to acquire that skills they should be equipped with sensors which can detect the visible or infrared spectrum of light. Understanding optics plays a significant role in developing computer vision based systems. If we are looking at sophisticated cases of machine vision, the understanding of quantum mechanics is also necessary. Also, the more we understand how the human vision works, the better we can develop a system which can replicate them. Therefore, an understanding of neurobiology related to human vision can also enhance our innovation levels.

Once we receive the required data from the sensors, the next step is to process that data and that when we rely on image processing and image analysis.

Applications of computer vision:

The inspection processes can be made much more comfortable using computer vision. If we are considering some hazardous work environment, we can just let the machine do the hard stuff for us. In places where identification of objects requires human intervention, these machines can be used to assist humans so that the result is obtained faster.

Navigation systems can be built around this like in the case of autonomous vehicles which rely on the traffic data which is in front of the vehicle.

In industries, the production can be controlled, and the process can be made more efficiency by making a machine overlook the entire process

When a simulation is required without damaging the original object, a model can be created using computer vision so that invasive tests can be simulated on the model.

This can be used for surveillance so that any undesirable activities can be kept under check

One of the best application will be in the medical field where a lot of images are used for diagnosis.

We can see how the computer vision fits into different sectors in the chapter dedicated to applications of the blockchain, ai and IOT.

Summary
- Computer vision deals with enabling machines to respond to the environment just like humans respond to vision.

10. Robotics

Preview
- Basics of Robotics
- Attributes of Robotics

There is an endless number of things to discover about robotics. A lot of it is just too fantastic for people to believe.
-Daniel H. Wilson

Robotics was the result when mechanical engineering, computer science, electronics, and others combined. It mainly deals with the design, manufacturing, working of robots as well as the systems which controls the robots. Karel Čapek introduced the term robot in his work Rossum's Universal Robots. In this play, a factory creates robots which could be mistaken for people.

There are robots developed for many specific tasks like the assembly line work and the robots with the capability of handling heavy units in industries.

What are the essential attributes associated with robots- The mechanical part, electrical part and the way it is programmed. The mechanical parts works based on the principles of real problem statements set by the environment. The mechanical part's goal is to accomplish a physical task like robotic arms used to carry heavy components in industries. What drives these robotic arms? Where do they get the power to operate? The robotic parts are powered by electricity. Therefore, they should be backed by an electrical circuit. The electrical parts are also responsible for collecting data using the sensors. How does the robot decide what to do? That's where the computer programming comes in; the robotic arm will be programmed to detect the object to be moved from the pickup point to the destination. Programming is the brain of the robot, which makes the full use of the resources available from mechanical and electrical components. The robots, when powered by artificial intelligence, can interact with the environment on their own and make decisions accordingly.

When the power source of the robot is decided, the attributes like safety, weight and the durability is taken into consideration. The battery can be even connected externally which will provide extra room, but the disadvantage here is that the robot will always be connected to an external entity. The commonly used power sources are solar, hydraulic, pneumatic, flywheel energy and nuclear energy. Just like the actuators discussed in the Internet of things, actuators are the moving parts in the robot. These are responsible for achieving the mechanical objectives. The actuators used in robots are Electric motors, Series elastic actuators, linear actuators, muscle wire, Electroactive polymers, Piezo motors, Elastic nanotubes.

The robots are expected to perform their activities based on the condition of its working environment. To understand the environment, sensors should be used to collect data about the environment. The sensors are used for real-time measurements and trigger safety warnings. The major sensing systems are touch and vision. The robotic arms used the same mechanism of sense as the human hands when the touch-enabled robot touches an object the fluid behind its surface deforms which can be used to profile the object based on the touch. Just like the touch, the vision sensing also works based on the basics of human vision; we have dedicated a chapter to "Computer vision," read that to get a better idea of the concept. Other sensing mechanisms include radar and sonar.

Robots deal with objects using effectors. The effectors are enabled using the mechanism set by the actuators. Effectors are responsible for producing an effect on an object which can be any modification or displacement. The typical examples are mechanical grippers and vacuum grippers. The movement of the robots can be enabled in many different ways like rolling, wheel balancing, multiple wheeled, tracked like tanks, walking, hopping like a pogo stick, flying, snaking, skating, climbing, swimming, sailing. The navigation system can be powered by a combination of radar, GPS and lidar which can be used for accessing the environment and to avoid obstacles.

There should be progress in some attributes to make the human-robot interactions more effective. Those attributes are Speech recognition, voice, gestures, expressions, emotions, personality, and social intelligence.

The robotic engineers are responsible for:

- Designing robots
- Maintenance
- Application development
- R&D to improve the spectrum of application

Humans either control a significant portion of the robots in today's industries, or they are working in a relatively static environment, with the development in AI sector, there will be more autonomous robots deployed in the industries.

Summary
- Robotics is an interdisciplinary field where Computer science, Mechanical Engineering, and Electronics combines.
- Safety, weight and the durability is taken into consideration while selecting a power source for the robot
- Touch and vision are the primary sensing mechanism in robotics.
- The Robots produce an effect on external objects using effectors.

11. Uncanny valley

Preview:
- How much should artificial intelligence resemble humans?
-

"Whenever I hear people saying AI is going to hurt people in the future I think, yeah, technology can generally always be used for good and bad, and you need to be careful about how you build it ... if you're arguing against AI, then you're arguing against safer cars that aren't going to have accidents, and you're arguing against being able to diagnose people better when they're sick."

-Mark Zuckerberg

Have you ever experienced an eerie, unsettling feeling when you are watching some entity closely resembles humans but not entirely similar? This is contradictorily comparing to how we feel more comfortable when objects get more human-like. If we plot a graph denoting a resemblance to humans against comfort levels, we can observe a dip or a valley in comfort, and this is the uncanny valley.

This phenomenon can we experienced in simulations, robots, animations, and humanoid toys. There are many theories which explain the cause of this cold feeling:
When a non-living entity looks almost like humans, it evokes a sensation of fear of death and deformity.
A feeling that, there is a possibility of us being a soulless creature is also fear-inducing.
These humanoids are replicas of real-life people which will induce fear of being replaced.
When the robots exhibit movements which are not human-like, there can be a fear of losing control over one's body.
Our cognitive system must have evolved to such a level that we can sense inferior immune system, hormonal activity, and fertility. The humanoid features can be a direct suggestion to these features, which will automatically evoke such a repulsive response.
Talking about cognition, the chances of this repulsion can also be due to our capacity to avoid disease-causing entities by evoking a sense of disgust.

So it is clear that we should be avoiding the possibility of creating robots with uncanny valley. How can this be accomplished?

The primary reason for this sensation is due to the mismatch between appearance and behavior of the object. When the robot is more human-like in appearance and more robotic in behavior, it will trigger the effect. The uncanny valley is also triggered when behavior is more human-like than the appearance. This same fact can be used to solve the uncanny valley; you have to balance the human nature in appearance and behavior.

Why does this mismatch trigger such an effect?
The reason is the expectation when the appearance is more human-like we expect human behavior and vice versa. This same effect can also be observed when there is no balance between face texture and face proportion. To avoid the effect, it should be made sure that the face proportion matches the face texture.

Summary:
- When several attributes of artificial intelligence are considered, if an attribute resembles humans greatly than the others, there will be a discomfort evoked known as the uncanny valley.

12. Major Companies and their AI initiatives

"If you had all of the world's information directly attached to your brain or an artificial brain that was smarter than your brain, you'd be better off."
— Sergey Brin

Most of the prominent technology companies understand the potential of Artificial Intelligence Market, and they are competing with each other to capture this market. Let us have a look at some of the companies and their offerings in this segment.

Watson is a cognitive system developed by IBM which enables a new partnership between humans and technology. Initially, it was created as a question answering computing system that IBM built to apply advanced natural language processing, information retrieval, knowledge representation, automated reasoning, and machine learning technologies. Today, Watson can understand different forms of data, learn them and interact naturally with people. There are three pillars for a Watson system to work: Data, information, and expertise.

So what is Watson explorer? With the explosion of data, the amount of data being generated is tremendously increasing day to day a lot of them being unstructured. Watson explorer is a content analytics tool for enterprises, which accepts all forms of data inputs, structured or unstructured from different systems in an enterprise and presents them to the users in a single view. Further, Watson explorer delivers data, analytics and cognitive insights which is relevant to a particular employee through its 360-degree information application. This significantly helps enterprises to reduce the time wasted over information search.

The content analytics platform of Watson explorer helps enterprises to derive insights from different data such as customer feedbacks, research reports, production data, etc. and to use those insights to streamline business processes and strategies.

Thus regardless of format or location IBM Watson mines data to derive patterns and insights and lets users to explore the information which is relevant to them. Listed below are some of the use cases were IBM Watson explorer content analytics caters to.

Customer Insight
- Customer experience
- Customer satisfaction and survey analysis

- Product and service quality
- Churn prediction
- Marketing campaign development and execution
- New revenue opportunities
- Product enhancements

Crime Analytics
- Community policing
- Investigation analytics
- Incident management
- Antiterrorism initiatives
- Cybercrime investigation

Healthcare
- Diagnostic assistance
- Clinical treatment
- Critical care intervention
- Research for improved disease management
- Fraud detection and prevention
- Voice of the patient
- Claims management
- Prevention of readmissions
- Patient discharge and follow-up care

Insurance
- Risk assessment
- Fraud detection
- Policy and underwriting analysis
- Claims analysis, payment validation and loss review
- Reserve trending and optimization

Finance
- Anti-money laundering
- Internet banking fraud
- Operational efficiency
- Risk management and compliance

You must have come across the expression "Hey Siri." SIRI is an intelligent personal assistant which Apple uses in its different devices. SIRI lets you use your voice to do a variety of activities such as send messages, make appointments, control the apps in your phone, and update your calendar and to-do list. SIRI has a speech recognition engine powered by advanced machine learning techniques which include recurrent neural networks.

You must have seen the black cylindrical box sold by Amazon known as Alexa. Alexa is a voice assistant developed by Amazon for certain Amazon devices and some other third party devices certified by Amazon. Alexa can be used to instantly play music, control your smart home and provide information, news and much more just using your voice.

13. Will AI steal jobs?

Preview:
- Can AI replace humans in jobs like driving?

"It seems probable that once the machine thinking method has started, it would not take long to outstrip our feeble powers... They would be able to converse with each other to sharpen their wits. At some stage, therefore, we should have to expect the machines to take control."
                                            -Alan Turing

There is a widespread notion that when the mainstream adoption of artificial intelligence, human intervention will be reduced to a large extent which will make people lose jobs. This is when I'll be giving you a glass; I'll slowly pour water into it till the halfway mark, and I ask you to describe the glass. If you believed that AI is going to replace humans and people will lose jobs, well the chances are that you will be telling me that the glass is half empty. In short, seeing this as a job loss event is a pessimistic approach. What about the endless possibilities that can emerge when AI is working for us?

Imagine the condition of Assembly line workers; they are mostly working around heavy machinery. This repetitive and physically taxing process can expose these workers to injuries. Why not let the AI do the job?

Another repetitive and monotonous work is that of the driver, even though this requires a remarkable breakthrough for implementation because of the scale of people this change can affect. Imagine the position of a taxi driver who is going through the repetitive process and facing the strains of possible traffic issues. The journey is predefined in this case, the automation of drivers looks easy to achieve on paper, Unlike automation in industries, the scale of adoption that this requires is what makes it hard.

Extraction of minerals mostly include working in harsh conditions, why not let the AI do the hard work and let us oversee the process and make sure everything is going as planned. Mining coal from underground mining locations includes many risks like the collapse of the mining location and the chances of being exposed to harmful and toxic gases. If AI can be used in this kind of environments, the process just has to be supervised and made sure AI is functioning as required, with AI applications emerging in every sector there will be endless job opportunities as well as endless training programs for people to make them ready for occupying job positions, so isn't introduction of AI to do the risky and physically taxing work is a win-win situation, win for the workers as well as win for the final product as the process can get more controllable and predictable with the use of AI.

If AI replaces these jobs, the people will get exposure to better job opportunities which will increase their living condition by several folds. Everyone will be shifted to a sort of managerial role rather than getting involved manually in the process. This automation process will increase the productivity which means work hours will be reduced. We often hear about the stress levels related to the working hours and intensity of work. We will get more time for ourselves which will mean the practical situation will become a liability. When AI is taking away the work intensity and work hours, the result will be more satisfaction and happiness, so this seems like a good trade-off when we are comparing it with the temporary job loss situation.

Summary
- AI will replace trivial as well as dangerous jobs
- The replacement will give rise to new opportunities in the field. Therefore, overall living conditions will get superior.

14. Ethics of AI

Preview:
- The level to which AI can replace humans
- The disadvantages of adopting AI
- Setting basic protocols for AI

"I visualize a time when we will be to robots what dogs are to humans, and I'm rooting for the machines."
-Claude Shannon

When did you first experience a surge of ethical concern related to machines? Maybe it was while you were watching the movie "Matrix." You might have imagined yourself in the shoes of the lead character Neo. You are stuck in your daily routine- work, eat, sleep and repeat. Suddenly one day, you are approached by someone or a group of people offering you to show the "reality." Now, you decide to take their offer after a great deal of dilemma. Next moment when you wake up, you realize your entire life that you remember was a big lie- a simulation, a made-up world. How right do you think that is? This scenario raises a question in the community about what machines could do in the future and what we can do so that we are not suppressed by machines as shown in the movie.

Ethics, also known as moral philosophy is the study and resolution of what's right and wrong. The resolution is made by defining the positive and negative entities like good and evil.
Even if we are not considering this extreme dystopian situation, there are other questions which can be raised. These are the things with higher chances of happening shortly. Think about what you feel regarding the following questions

Should AI be used in warfare? If it can be, to what extent?

Should AI replace customer service interactions?

What about the crucial positions like judge, soldiers and other law enforcers?

According to Joseph Weizenbaum, these kinds of positions require human empathy to function correctly. These positions demand the essence of respect and care to operate. Consider a bedridden person, imagine how much difference it would make if an artificial being is taking care of him compared to when a human being is doing the same. We demand empathy directly or indirectly; we are hardwired for that.

Once you remove the empathy part by replacing humans with machines, there will be chaos. It is said we humans can recognize patterns which machines fail to, these activities happen more on a subconscious level, and it mostly reflects as intuition. We perceive this world through our five senses and our ability to replicate human behavior is restricted to the five senses. If human interactions involve the higher level of sense and this artificial can't deliver that, we will find ourselves trapped in a demoralized and frustrated society.

According to Bill Hibbard, due to the immense impact that AI can create on humans, he is suggesting that all the resources related to ai should be kept transparent. This step can be seen as a great leap towards better ethics in the artificial intelligence sector. Elon Musk created a non-profit AI open source research company known as OpenAI. The other open-source AI projects include OpenCog which is an open source framework for developing AI.

What will happen if intelligent machines are used in warfare?
When these machines are acting autonomously, can this lead to a Skynet situation as shown in terminator? When these war machines are getting more advanced, apparently it will be imparted more and more autonomous behavior. According to the researchers, these machines will show more human nature due to increasing free nature. When robots are assigned responsibilities, it should be made sure it won't backfire by setting and programming morals and ethics in them. These machines should never deviate from the moral baseline, which will ensure the safety of humanity.

If an ethical baseline is set, the robots can develop a logic to decide their actions which might be quite the opposite of what we desire from the robot's actions. It will be just like a soldier going rogue; the only difference is that this soldier will be more efficient and better equipped.

Artificial intelligence can make our life more comfortable that's for sure. AI can find a better cure for diseases, and they can help the society function more efficiently. The chances of severe drawbacks regarding research on AI for using it in warfares could not be ignored. We don't want to end up creating something which we are unable to control. In Jan 2015, AI experts like Stephen Hawking, Elon Musk signed an open letter on artificial intelligence which was aimed at extracting all the benefits of AI, without compromising the normal functioning of the society.

This open petition raised concerns like, should autonomous vehicles be banned. The reason being, there can be situations where these machines have to decide between a high chance of a small accident and a small chance of a major one. The problem here is, how is a high accident differentiated from a smaller one. This can end up being a very biased situation. Already many autonomous vehicles are being successfully deployed, so we can see that this is a short-term concern of the petition. The long-term concern includes the previously discussed AI making decisions which can compromise humanity.

Nick Bostrom is suggesting that artificial intelligence can make the human race extinct in his work "Ethical Issues in Advanced Artificial Intelligence." According to the author, the AI will be capable of making its own decisions. It will have the capability of designing its own goals and also will possess the capacity to make plans to achieve those goals. Using this level of higher independent thinking, it can even influence other machines to function to achieve its self-designed objectives or even pull the plugs of the machine which opposes its initiative. Considering these negative aspects, we can't stop making progress in artificial intelligence sector because the positive impacts look very compelling. The only solution is to find a balance, by setting proper moral and ethical baselines which the robots can't deviate from.

Isaac Asimov showed through his work I, Robot that no matter how we set the moral baseline, there will always be unexpected and unanticipated behavior. According to him however the laws are set, it is just a matter of time that there will be circumstances which will break it. Therefore, the future of AI dramatically depends upon the magnitude to which a reliable, ethical system can be integrated into the machines.

Summary:
- AI can't replace humans where empathy is required.
- AI initiatives shouldn't backfire by giving AI the high capability to make decisions on its own.
- There should be a balance between progress in AI and the basic guidelines which AI should follow to prevent any undesired behavior.

BLOCKCHAIN GUIDE

1. Byzantine Generals

Preview:
- What is the Byzantine Generals problem?
- How to solve the Byzantine problem in distributed networks?
- Solution to lack of trust in a distributed network

"I don't need a hard disk in my computer if I can get to the server faster... carrying around these non-connected computers is Byzantine by comparison."
                                        -Steve Jobs

If we want to know the importance of Distributed Ledger Technologies, first let's understand the Byzantine Generals Problem.

Byzantine army decides to invade a city. The army is headed by a group of generals, among whom the army is distributed. The city has to be surrounded by the army with the generals leading their fraction of army. The generals can only communicate with each other using messengers. For the attack to be successful, the attack has to be launched simultaneously from all directions. If the decision is to retreat, they have to retreat together. If the actions happen out of sync, the army will lose the battle. If a portion of the army attacks and the other portion retreats, the Byzantine army will lose terribly. Therefore, any decision made should be synchronous throughout the army.

When all the generals and messengers are loyal, this is a very healthy situation. In this case, if a plan stating a launch of attack at midnight is passed on, the messengers will successfully deliver the message and the attack will be successfully initiated at midnight and the city will be invaded with ease. Even if there is a difference in opinion among this loyal bunch of people, they can still conclude by voting on the topic and by acting on the popular opinion whether it is attack or retreat.

Things aren't this easy if the loyalty of the generals is questionable. A rogue general can pass on wrong information which will be conceived as the right information. Things get complicated when information is wearing the cloak of righteousness. The army will be split because of the wrong information passed on by the rogue general. A part of the army will attack the city while the other retreats, this event will cause a terrible loss to the army and they will lose horribly. The same situation can happen if a messenger is turning against the army and passing along wrong information.

Suppose there is voting among the generals to decide whether to attack or retreat and if the final deciding vote lies in the hands of this treacherous general. He can send messages supporting the action to both the parties, which will create the same situation as just mentioned- a mixup of attack and retreat.

We can see that there are a lot of points of vulnerability in this process of communicating the message. Whenever a single point in this network of communication fails whether the general or the messenger, the plan is doomed to fail. You must have understood that there is one thing that is required for the successful invasion of the city - TRUST. Without trust in this system, it is doomed to fail. This situation applies to every case involving a network with multiple decision making or confirming points(nodes). What if the requirement of trust was removed from this issue? Then we no longer have to worry about trust.

Distributed Ledger Technology accomplishes this feat by creating a record which is shared with everyone in the network. This record is immutable, all the recorded data in this ledger cannot be modified. If you want to add a new data to this network, it has to be approved upon by the majority of the network. If Byzantine army used Distributed Ledger Technology, when an attack order is issued by a general it will be visible to all the generals in the network, and the process will be approved if they reach a majority consensus on the decision. It can be seen in the record if the decision has been approved or rejected. Based on the decision reached in the record, they can act simultaneously understand whether it is attack or retreat. See, how magically trust was removed from this act and everything works just fine.

The bitcoin solves Byzantine failure by using proof of work algorithm, which is used to correctly guess the hash function associated with a transaction which will be verified by the entire network after a node guesses it correctly and thus preventing any double spending issue. We will get into details of that in the chapter dedicated to bitcoin.

If we were trying to solve this issue by increasing trust, the effort would have been futile as humans are prone to error. The solution was hidden in plain sight- remove the trust from the system.

Summary
- Byzantine problem generalizes a situation where there is lack of trust between the participants in a distributed network.
- The blockchain is the solution to this problem under its permanent, transparent and secure nature.

2. How blockchain works

Preview:
- Origin of blockchain
- Understanding a blockchain network

The old question 'Is it in the database?' will be replaced by 'Is it on the blockchain?'
-William Mougayar

The idea of Blockchain was proposed by Satoshi Nakamoto in 2008. Using this technology he developed a peer to peer electronic cash transaction system known as bitcoin which solved the issue of double spending without any involvement of a third party. Bitcoin played a significant role in introducing the concept of Blockchain to the world.

Blockchain has a very high Byzantine fault tolerance because of its inherent feature of accepting modifications after achieving mutual consensus from the entire network. The blockchain is suitable for all the situations where trust is required among the participants and the data recorded is not to be manipulated.

Just like the name suggests, it is a chain of many of many blocks which grow over time. These blocks contain the details of transactions in the network, the timestamp corresponding to the transaction and the hash which links it to the previous block in the blockchain. When a group of data like transaction details is fed into a cryptographic hash function, it gives an output which is a much simpler version known as the hash. The hash is used to identify each transaction as is it unique to every transaction.

Every participant in the network has a copy of the blockchain record, any addition made in the blockchain requires majority approval from the participants. If the mutual consensus if not met, the new data will be rejected and if mutual consensus is reached, a new will be added to the blockchain which will contain all the details of the transactions - the public addresses of the participants, timestamp as well as the amount of currency involved which is case of a peer to peer cash transaction network. The data once added to the blockchain cannot be modified or deleted. The immutability feature is inherited by every blockchain network.

In a blockchain network, the authority doesn't rest with a centralized system, but it is distributed throughout the system. This decentralized nature of the blockchain increases the security of the system since there is no single point of vulnerability. If we are considering a bank which enables transactions, if the bank servers are hacked, the entire network depending on that bank for transactions will be compromised because the bank is a centralized entity and only has a single vulnerability point which cannot be compromised. In case of a decentralized network, the system won't be compromised unless you take control of the majority of the nodes in the system which is very difficult to achieve in a real life scenario.

The first mention of this structure of using blocks and chains were suggested by Scott Stornetta and Stuart Haber in 1991. The aim was to create a network which cannot be tampered with or manipulated once the data is time-stamped and secured. A year later they introduced the Merkel tree into this structure and thus improving its efficiency. It was 17 years later that this idea was popularized by Satoshi Nakamoto using bitcoin. As mentioned earlier the size of the blockchain file keeps on increasing with each new file being added to it. The bitcoin blockchain size reached a massive 100GB by 2017.

Bitcoin was just a start, a single application of the concept, it was just the tip of the iceberg. What mattered was what was hidden under the water- the broad spectrum of blockchain applications. This phase was called blockchain 2.0 where the concept of blockchain could be used for producing decentralized applications.

Blockchain can enable the exchange of everything which has value more comfortable, faster, cheaper and secure. Smart contracts can make sure that there is a fair exchange and the permanent nature of the records will make auditing much more comfortable. Blockchain also eliminates the possibility of double spending by using algorithms like proof of work, which confirms the hash related to the transactions. The hash associated with the transactions are responsible are linking the blocks together in the network. The longer the blocks gets, the more secure the network becomes since each new block reaffirms the validity of every other block in the network. If multiple blocks are created in the networks, the longer one will be sustained while the other is eliminated. The eliminated blocks are known as orphan blocks. Due to the ever-growing nature of the blockchain, in case of any discrepancy always the branch with more blocks will be preferred, which is the one with most mutual consensus run on it and thus more secure. The time taken to generate a block in the network is known as block time since the validation of transactions depends on the process of block creation; lesser the block time faster will be the transaction.

Summary
- The blockchain is a digital ledger which keeps a record of all the transactions in a network using cryptographic algorithms.
- The authority in a blockchain network is distributed throughout the network.
- Blockchain avoids double spending using consensus mechanism like proof of work.

3. Features of blockchain

Preview:
- Decentralized nature of blockchain
- Potential to replace the current system

"A lot of people automatically dismiss e-currency as a lost cause because of all the companies that failed since the 1990's. I hope it's obvious it was only the centrally controlled nature of those systems that doomed them. I think this is the first time we're trying a decentralized, non-trust-based system."
—Satoshi Nakamoto

In a network, the data can be spread across the network instead of storing it on a single system. It is a great challenge to manage such distributed networks which are made easier by the possibility of involving cloud computing which can revolutionize the distributed networks. Message passing is used for communication between objects in different systems. Traditionally for an action to be executed a function call is required which directly refers to the function or the process to execute it. Message passing sends the message like the name suggests so that the action is executed when it reaches the object in another system. This Distributed network and the message passing system is the essence of a decentralized blockchain network.

This spread of data ensures that the risk faced by the centralized network is averted. Centralized networks have only a single point of failure. If the hackers manage to find a way into that single point, the entire network is compromised. In a blockchain network, each node in the network is encrypted by the combination of public key and private key which are required to access the account and thus providing the network an extra layer of security. The private keys are like passwords to access the nodes which are kept in secrecy. The only way the hackers can compromise the network is to own computational power which can exceed the total computational power of the entire network which is a complicated process unless you have a lot of resources at your disposal.

The ledger is stored in every node involved in the blockchain network. Therefore, there is a continuous process of sharing data and validation. The message passing is done on a best effort basis, which does not guarantee the quality of messages or that the message is delivered. Best effort to deliver the message will be made every time a message is initiated. When a node successfully mines a block, this message is distributed across the network, and everyone else validates the block and adds it to the network, the block is added with a timestamp using algorithms like Proof of work like that of bitcoin.

Most of the blockchain networks are open source enabling anyone to contribute to the network or to use the source code of the network to build new blockchain networks. Any change made to the blockchain should meet mutual consensus otherwise the community will be split, and there can be forking, which will be discussed in an exclusive chapter. The blockchain networks which anyone can access are known as permissionless blockchain or public blockchain. Private blockchains are the one with limited access to the network and the people only with access can use the network. These private networks do not depend on the validation of transaction like public ones; there is no benefit of validation in this case. From this fact, we understand that the private blockchains are essentially a centralized system with the power of the network resting on a central authority. There are no rewards involved in mining in a private blockchain because of its centralized nature.

Bitcoin has never been hacked, don't get confused with the hacks in bitcoin exchanges which are centralized bodies and are naturally prone to attack. To shut down a blockchain system, you can't just attack a single point of vulnerability and expect it to haywire like in the case of a centralized network. These nodes will keep providing consensus even if some of them are shut down by hackers since the message delivery is by the best effort model, the process will go on no matter what.

Traditional banks can take a great deal of time to process transactions whereas blockchain networks can enable the same transaction in much less time. This will also reduce the transaction fees involved in the process because the effort is minimal in this case.

Traditional currencies are not backed by anything, and it has value based on supply and demand as well as because the government decides so. The traditional currency has a lot of weak spots like being vulnerable to inflation. This problem is sorted by cryptocurrencies backed by blockchain network. The creation of cryptocurrency is backed by the computational power which is used to validate transactions in the network. There are several built-in features which prevent inflation like the self-adjusting mining difficulty, halving effect and the maximum cap on the number of coins that can exist.

Summary
- Centralized networks can be compromised because of a single point of failure.
- Blockchain stores the encrypted ledger in every node so that network can't be compromised when a single node is attacked.
- Any changes in the network should be agreed upon by mutual consensus.
- Blockchain can reduce transaction time and fee.
- Blockchain can back the currency system.

4. Limitations of blockchain

Preview:
- How far can blockchain deliver what is promised?
- What are the problems blockchain networks are facing currently?
- Solutions to current Blockchain issues

"The difference between stupidity and genius is that genius has its limits."
-Albert Einstein

The blockchain is an entirely new technology which can rewire the working of industries; this rewiring will demand the people to acquire new skill sets to survive in the community. With the industries getting more and more efficient small jobs will be replaced which means the people carrying out those jobs will be forced to learn new skills. This is not a bad thing if we are thinking optimistically as their standard of living can increase with better jobs. The gap can be filled by the initiative which can educate the people more about emerging technologies.

As we well know blockchains functions by the activity of several nodes which enables the network. The underlying ideology of blockchain is to be a decentralized network, for this to happen, the number of nodes in the network should be large enough such that the initiative of a single entity cannot overthrow the network, by forcing the network to favor decisions on their side. Mainstream adoption of blockchain is a necessity for its proper functionality.

Scalability is an issue faced by many blockchain networks. The network's strength is determined by the capacity of the individual nodes. Each node in the network is expected to process every data or block that is added to the network. This will create scalability issues as there will be a limitation to how much a node can contribute. The solutions proposed for this issue are the implementation of the lightning network and sharding. The lightning network creates an offline sideline which will be updated instantaneously so that it can add it to the leading online chain later on as the capacity of the node permits. The sharding is the process of splitting the data among nodes in such a way that they can be recalled when they supposed to be reviewed or reused again.

The blockchain is an immutable network, so what can we do if there is a human error associated with an entry in the blockchain. If the solution is to enable edits, it will be going against the essential feature of blockchain to be immutable. The opportunity of mutation will compromise the security feature is added to the network. The possible solution is to enable direct entry of data from sensors (internet of things) so that no human intervention is necessary and even if there are changes required it should be adequately conveyed to the network participants.

As we discussed earlier, for the proper functioning of the blockchain network, the nodes should be spread on a large scale. If the use of blockchain network is confined, the chances of the network being invaded are more. This already happened with the bitmains mining pool when they crossed the 51% dominance in the bitcoin blockchain network. The issue was solved when they agreed to reduce their participation to around 30% deliberately.

Summary
- There is a high requirement to educate people about the new technology for the progress to happen.
- Proper decentralization can be achieved only by large-scale adoption
- Blockchain currently faces scalability issues, one of the solutions is implementing the lightning network.
- Entries in blockchain ledger should be perfect as the record is immutable.

5. Public blockchain and private blockchain

Preview:
- What is the difference between a private and public blockchain?
- Should we use a private or public blockchain network?

"Ten percent of my net worth is in this space,"
–Mike Novogratz, hedge fund manager, Galaxy Digital Assets

The public blockchain is open to the public just as the name suggests; anyone can join the network without any prerequisites. The only requirement is to have an internet connection and software to enable connection to the public blockchain network. Anyone can validate the transactions in the network and can maintain the record of transactions in the network. The best example of public blockchain is the one who popularised the idea of blockchain- The bitcoin. Anyone can join the bitcoin network, use their processing power to contribute to the network for validating transactions and mine bitcoins. When the number of participants is increasing in a public blockchain, it will be stronger and more secure. More participants or nodes in the network will result in better validation of the transactions. It will be tough to compete against the processing power of the network when it has been adopted mainstream which will ensure that the system will remain decentralized.

Unlike Public Blockchain networks, the private ones require an invitation from the existing member. The entry is mostly verified by the guidelines set by the network. The entry is restricted based on the requirement of the network. Only after the entry to the network has been accepted, a ledger of the network can be maintained. Since this is a controlled environment, the power will rest will the people who created or are responsible for the network. This is necessarily a centralized system with authority having final say on the decision of the network. Even though the network is centralized and it does not follow the fundamental revolutionary ideologies, it has a wide range of application which will make industries more efficient. Using private blockchain, you can tokenize every activity in a particular area if you are using a particular gym which has enabled a private blockchain which rewards you based on the effort you put in for working out. This gym lets you redeem this token for various offers at the gym; this application will be confined inside the gym. If you want to join this network, gym membership will be an essential condition, and the tokens will not have any use outside of the gym. The guidelines of the token generation and token usage are determined by the gym owners, which we cannot vote against. This is now private blockchain works.

Public blockchains are more secure than private due to the decentralization factor. Private blockchain can get compromised if the primary system controlling the activity of the network is hacked into, this is impossible in the case of public blockchain where it is quite difficult to impossible to get control of more than 51% of the computational power, which btw is the minimum requirement for hacking the network. Less computational power is required for maintaining consensus in the private blockchain networks and therefore is more eco-friendly.

Summary:
- You can select the nature of blockchain network you want to use depending on whether you want to keep the network entry open or restricted
- The public blockchain is used for an open network and private for the one with restricted entry.

6. BITCOIN

a. A Bitcoin transaction

Preview:
- When should you use bitcoin?
- How can you use bitcoin for cash transaction?
- How can you store your bitcoins?
- What decides the value of bitcoin?

"Well, I think it is working. There may be other currencies like it that may be even better. But in the meantime, there's a big industry around Bitcoin.— People have made fortunes off Bitcoin, some have lost money. It is volatile, but people make money off of volatility too."
                                        -Richard Branson, Founder of Virgin Galactic

Imagine your friend Arnold is contacting you from Africa, he needs $10,000 as soon as possible. You are a wealthy businessman, and you quickly have that kind of money at your disposal. You want to send this money directly without any involvement of a 3rd party like banks or other financial agencies because you don't trust the security offered by these kinds of centralized organizations. What option do you have now? - The Bitcoin. After 2009, Bitcoin emerged which can enable peer to peer electronic cash transactions without the involvement of a 3rd party. To transfer the money all you need is a group of letters known as the Public address of the Bitcoin wallet which is unique to every person owning bitcoins.

Now you have figured out how to send the money, which is by using bitcoin. You even asked Arnold his public address, which he had already created beforehand. The public key is like the account name in every social media. How do you use account names in social media? If you want to find someone, you can easily find them by using their account name, and if you want to send a message to those people, you identify them with their account name before sending the message. The public key is the similar feature to account name in cryptocurrency world. There are lots of difference in how it functions as compared to account names.

You cannot search up public keys as you do with account names. This is because of the semi-anonymous nature of bitcoin; we will get to that in detail soon. If you need someone's public key, you have to get it directly from them; there is no option to search it up. If there is an account name, there should be a password right? Here the Private key is the password. You access your bitcoin wallet using your private key. The private key is the heart and soul, and it should never be lost or misplaced. If you recklessly place your private key somewhere and someone finds it, he or she can easily siphon all of your money, and you will have no means to figure out who did that.

You got Arnold's public key now, but wait do you have one? It is ok that we asked for his critical first because we'd be doing the same later on, to process the transaction. Bitcoin public key is associated with a Bitcoin wallet, so to have a public key, we have to set up a wallet. A wallet is essentially software which enables us to process transactions and check our bitcoin balance. The bitcoin wallet is a must for handling Bitcoins. When you hear wallets, if the image of traditional wallets is popping in your mind, well you are picturizing it wrong. A bitcoin wallet is essentially records of transactions stored in the bitcoin blockchain. Now, let us see the bitcoin wallet options you can consider.

Wallets can come in the form of software, hardware or even paper offering different levels of security. They can be free or paid, and the investment should be made according to the level of your cryptocurrency investment.

The first option is a Desktop Bitcoin wallet, and you can download the software which can be free or paid. This wallet only allows access from the system, where is it downloaded and setup. This exclusive nature can serve as a security feature. What if this computer gets hacked or is infected with malware, then you will lose all your Bitcoins. Malware which can track keyboard inputs can easily track your private key and sabotage your bitcoin treasure.

The next option is an Online wallet, which is managed by a third party website. On the plus side, it can be accessed from anywhere in the world because all you gotta do is log in to that website, an internet connection is all that is required. These wallets can be highly vulnerable since they are controlled by a centralized organization. If they are hacked, you will lose all of your bitcoins stored in your online wallets. The popular exchanges like Mt.Gox and Bithumb have been hacked which drained away millions of dollars from people.

You can use mobile wallets which functions just like any other app. This will be a very simplified version as compared to a desktop wallet since mobile phones run on lower specifications. Even mobile wallets face the threat of being connected to the internet and the risk of being infected with malware.

Now let's consider a safer option, Paper Wallet. Paper wallets are physical printouts which contain the public key and private key. This is not at all connected to the internet which makes it highly secure. The keys are mostly displayed as QR codes. During transactions, you have to use the software wallet to transfer the bitcoin, to process the transactions temporarily. You have to make sure you don't lose or misplace your paper wallet, and there is no means of recovering your account once you lose your paper wallet. If you are the kind of person continually misplacing stuff, reconsider buying a paper wallet.

We saved the best for the last- The Hardware wallets. This is the most preferred means of storing Bitcoins. Just like paper wallets, they are disconnected from the internet and therefore do not face the evils of the internet like malware and viruses. This looks like a fancy USB drive, and hardware wallets can be used by plugging into your USB ports. They can support a wide range of cryptocurrencies, and they are supported by many software and web wallets. Hardware wallets like Ledger Nano S and Trezor come with a LED which displays the balance and using which you can confirm the transaction using a unique pin code.

There is an account recovery option which comes along with the hardware wallet, a recovery sheet. You can choose a long string of words or letters which can be used to recover your account if you lose your wallet. This is a risky option because if you lose that recovery passcode, anyone who finds it can access your account. Even though Hardware wallets are on a higher side of investment, they are worth for what they offer.

Considering all the options above, now you decide to buy a hardware wallet but when you find that it will take at least a week for it to be delivered due to high demand. Anyways you order one since you are a great believer in decentralized currencies and you plan to use them on a regular basis in future. Arnold needs his money A.S.A.P, he can't wait for a week, so you decide to set up an online wallet. Most of the online wallets will require your KYC(identification), this is to prevent illegal activities like money laundering and other kinds of illegal transactions. Now you upload all your KYC information and photos which are required, and you patiently wait. This can take up to a couple of days depending on the online wallets you are using. After the KYC is confirmed, after logging in you find an option to load the account with fiat currency(like US dollars). You now load the $10,000 which is required by Arnold. Most of the online wallets also have the option to buy Bitcoins along with storing bitcoins feature. Now you buy Bitcoins using that $10,000 and transfer those bitcoins to Arnold which he converts to his corresponding fiat currency or uses it as bitcoins itself depending on the nature of his urgent requirement.

The thing is most of the time Arnold won't receive $10,000, on some occasion it can become $13,000 or $6000. This is due to the volatile nature of the bitcoin. Bitcoin prices were highly fluctuating, in 2017 Bitcoin experienced massive fluctuations rising to $20,000 and dropping back to $13,000. Why is Bitcoin value so volatile?

Even though Bitcoin was released in 2009, it can be considered as a relatively new player to the financial sector. There is a lot of misinformation which is floating around bitcoin sugar coated with hype which sure does excite people but fails to impart any useful information. By the end of 2017, it is said that almost 95 percent of the currency was owned by almost 5 percent of the entire Bitcoin investing community. This gives those creamy layers, the power to manipulate the prices by processes like pump and dump.

When some regulations are in place which prevents or decrease the movement of bitcoins, there will be a marginal decrease in accessibility which will decrease the demand and thus the value of Bitcoin. The value of bitcoins dramatically depends upon the number and influence of the access points(like exchanges) which play a significant role in increasing the familiarity and accessibility and thus increasing the demand. When an access point is cut off from the network, it will inevitably affect the prices.

A decrease in value can also be triggered when trust issues kick in, one of the most popular exchange, Mt. Gox was hacked twice, and the Korean exchange Bithumb was recently hacked which cost millions to the investors. When this happens there will be a marginal decrease in value of bitcoin and people will get more reluctant to use an exchange. To reduce these kinds of risks you can use exchange when you want to exchange your Bitcoins or buy bitcoins, otherwise, keep your Bitcoins stored safely in cold storages like hardware wallets.

There are a lot of internal issues within the bitcoin developing community due to the lack of governance. These internal issues can hinder the progress of the network and also raise questions of the developing community being stable thus reducing the value of bitcoin.

Bitcoin network will get more centralized when mining is facilitated majorly by a source of hardware like Bitmain(We will explore bitcoin mining in detail soon). They control the network by their control over this hardware and its ever-increasing demand. Crowd speculation plays a significant role when the price of bitcoin is intentionally manipulated because a high volume of investments depends on how bitcoin is moving, up or down. Despite all this volatility issue, when Bitcoin achieves more mainstream adoption, it can get more stabilized.

The transaction will work out just fine if Arnold wants to use the currency in the bitcoins. If he wants to use it as fiat currency and the value of bitcoins drop at that time, then the scenario will surely become complicated.

If the value of Bitcoin gets higher, Arnold can cash out the $10,000 and keep the rest of the currency in the bitcoins as a valuable investment, or he can exchange these bitcoins for other promising cryptocurrencies. The only thing he should take care of is that, after transactions are done, he should store the bitcoins safely preferably in a hardware wallet.

Now Arnold is happy having received with the cash he urgently required. After you first successful bitcoin transaction you are getting more interested in how bitcoin works and what is happening behind the scenes of these transactions. The transaction which you just experienced took almost an hour to process, and you might also be wondering if there are any similar processes which can offer better options.

Summary
- Involvement of third-party financial agencies can be avoided by using cryptocurrencies like Bitcoin.
- You can store your bitcoins in wallets; the most preferable and safe one is hardware wallet.
- Bitcoin's value can be influenced by the adoption of industries, the influence of bitcoin exchanges, media's stance on bitcoin and internal issues in developing community.
- Bitcoin can get more decentralized when it is increasingly adopted.
- The value of bitcoin can fluctuate significantly during a transaction.

b. Satoshi Nakamoto

Preview
- Introduction of bitcoin by Satoshi Nakamoto
- Who is Satoshi Nakamoto

"[Bitcoin] is a remarkable cryptographic achievement... The ability to create something which is not duplicable in the digital world has enormous value. Lot's of people will build businesses on top of that."
Eric Schmidt

When was the first time you heard about bitcoin? Was it the early stages of Bitcoin or was it when the whole thing became a phenomenon and started hyping up. Most of you must have googled a few common things all along the path after you were enlightened with the knowledge of the existence of bitcoin - What is the current value of bitcoin? Who created Bitcoin?

The question of bitcoin's value was always answered even though it changes from time to time, there was always a specific answer to that but, who created this worldwide phenomenon? If you can be satisfied with a plain answer, then here you go - Satoshi Nakamoto. That is all you can know about the identity of the creator. The identity is so unknown that people don't even know if it refers to a person or a group of people.

"Bitcoin: A Peer to Peer Electronic Cash System" was the first description of Bitcoin, it was released on October 2008 by Nakamoto on a cryptography mailing list known as metzdowd.com THe first bitcoins and the associated software to launch the network were released in January 2009. Nakamoto stated that the work on cryptocurrency began as early as 2007 and the Bitcoin was designed in a way that broad ranges of transactions are supported. Till around July 2010, Nakamoto handled the Bitcoin-related software along with other developers using the website, bitcoin.org. After this, he withdrew himself from the project and handed over everything related to bitcoin like source code repository and network alert key to Gavin Anderson.

When bitcoin was created, the mining was done by only Nakamoto. The first block, known as genesis block of bitcoin or the block number 0 had the value of 50 Bitcoins. The following text was included in the transaction description.

The Times 03/Jan/2009 Chancellor on the brink of the second bailout for banks.

The purpose of the sentence was to mark the date of the launch. We can see that the specialty of this block is that unlike the block which follows it, the genesis block doesn't have any previous block to link to. For the first 10 days after the launch of Bitcoin, Nakamoto was the only miner, and these Bitcoins are still unspent. To understand the magnitude of the value of this much Bitcoins digests the fact that during December 2017, Satoshi Nakamoto was the 44th richest person in the whole world. This was the period when bitcoin hit its peak value of almost $20k.

If we are going full-on Sherlock Holmes mode on Satoshi Nakamoto, we can see that this particular person or group has an affinity for using British English powerful words like "Bloody HELL." The first transaction's description also indicates that Nakamoto was inclined towards the London's The Times at the time of its release. From his bitcoin discussion forum Bitcoin.org, we can also observe that the person had strong command in English and the codes were not labeled in Japanese which contradicts the theories that he might be of Japanese origin.

Which the minimal information available, these are the people rumored to be Satoshi Nakamoto:

- Dorian Nakamoto
- Craig Wright
- Nick Szabo
- Hal Finney

Summary:

- Satoshi Nakamoto introduced Bitcoin to the world in the year 2008 through his White Paper, "Bitcoin: A Peer to Peer Electronic Cash System."
- There are a lot of speculations about the original identity of Satoshi Nakamoto, and it remains a mystery to the world.

c. How Bitcoin works

Preview:
- What is the role of banks?
- Avoiding double spending without the help of banks
- What is Bitcoin mining?

"PayPal had these goals of creating a new currency. We failed at that, and we just created a new payment system. I think Bitcoin has succeeded at the level of a new currency, but the payment system is somewhat lacking. It's tough to use, and that's the big challenge on the Bitcoin side."
-Peter Thiel, co-founder PayPal

We rely on third-party financial institutions like banks to perform transactions as we require a trusted intermediary who can execute and validate these transactions on our behalf. But have you ever thought about the level of trust we can have on these centralized banks? Have you ever thought about the errors that can happen with traditional banking system? When a centralized institution is entrusted with such an activity, there is always a scope of error which cannot be eliminated entirely. These errors are inherited when a third party is involved and will cause mediating disputes. The transactions records will be mutable since this centralized system is prone to error. We are relying upon these financial institutions to avoid issues like double spending. It is made sure that the money is not double spend by making sure it passes through the mediator in every transaction.

How can we avoid the mediator and still prove that the money is not double spend?
For solving this issue, we have to make sure that the previous owners did not try to process any other transactions. This can be ensured by maintaining a record of previous transactions. This record or ledger will bring awareness of the entire history of the transaction which will confirm that there was no attempt to double spend the money or bitcoin as in this case. To enable this system, this transaction record should be public, and each time it is amended, the majority of the network should approve the transaction. A timestamp is including with each new data that is added to the record.

The validating of the transactions are accomplished in the Bitcoin network by using the Proof of Work algorithm, to be specific SHA-256. Each transaction in the Bitcoin network consists of public addresses of the sender and the receiver as well as the transaction amount; these data are run through a hashing function. The hashing function generates a hash which is unique to every transaction. A transaction is said to be validated when the hash corresponding to the transaction is successfully discovered by the transaction validators in the network.

The people involved in validating transactions in the network are known as miners, and the process of validating the transaction is Mining. We associate mining traditionally with resources which are difficult to procure, so this is the digital analogy of the physical mining process. The miners in the network use the computational power of their system to find the hash function associated with the transaction.

For this mining process to happen, when a transaction is being processed, all the miners are alerted about the transactions by broadcasting the transaction throughout the entire network. The miners now try to find the corresponding hash function of the transaction. When one of the miners successfully finds the hash, the solution is again distributed throughout the network so that it is verified throughout the network and confirmed that the transaction is not a case of double spending. When the miners start to work on the next transaction, it is implied that they have accepted this hash. Thus, each point in the network which contribute to validating transactions is known as a node. Bitcoin network is essentially a network of nodes which actively validates transactions.

In the case of traditional currencies, they are issued by financial institutions. There is no centralized authority here to issue bitcoins, so the process of mining creates the new coins. This raises some questions like isn't mining the process of confirming transactions? How is bitcoin generated while mining?

When you are trying to guess the hash function associated with a transaction, it will consume a lot of computational power(high electricity consumption and high initial investment for a processor which can deliver high computational power). Why would someone invest so much, if he or she is not adequately rewarded or given a chance of being rewarded for his or her investment? That is why the miners are rewarded with bitcoins each time they successfully confirm a transaction and thus mine a new block. The reward system keeps the network running and growing.

Does this mean that everyone can go on investing in processors and earn bitcoins endlessly?

NO, The maximum number of bitcoins that can exist is limited to 21 Million, and this number will be achieved by around 2140 at the current rate. The Bitcoin mining algorithm is designed in such a way that the mining difficulty self-adjusts every two weeks. This happens in such a way that the average time for mining a bitcoin block is 10min. Each time a node mines a block, it will be rewarded with 12.5 Bitcoins and this reward halves every 4 years.

The final aim of the bitcoin mining algorithm is to adjust the block mining time to an average of 10mins. If more people are joining in the bitcoin network and the existing miners are investing more in mining processors, there will be an increase in computational power provided to the network. In this case, when the 2-week mark is reached the network increases the difficulty of mining suchthat the 10 min block time is satisfied.

What if miners leave the community and the processing power imparted to the network decreases? This will trigger the algorithm to reduce the mining difficulty such that blocks are mined in 10 mins. This mechanism will ensure the regular activity in the community. Whenever people are leaving the network, it will incentivize the people who stay more. When people see the increasing rewards, they will compel to join the network. This self-adjusting feature ensures the steady working of the bitcoin network, and it won't be running out of miners.

What will happen when 21 Million bitcoins are mined?
There are no more bitcoins to be mined now, very last of them is exhausted. The miners contributed to the network because they were rewarded, so happens now? There will be transaction fees associated with the transactions which will be rewarded to the miners to keep them motivated to provide enough computational power to the network.

The way that bitcoin works require no trust among the people involved, this is a trustless network, unlike the traditional money transaction systems. If we are considering fiat currencies like USD, they are not backed by anything but the value is just based on supply and demand. Fiat currency possesses value because the government is involved in maintaining that value. This is not the case with Bitcoin; it is backed by the computational power which was used to create it.

Summary:

- There are flaws in the traditional banking system like the issue of double spending and higher transaction fees.
- Bitcoin network makes use of SHA 256 Proof of Work algorithm to validate the transactions.
- The transactions in a bitcoin network are validated by miners for which they are rewarded with a certain amount of Bitcoins.
- The maximum amount of bitcoin that will be ever created is limited to 21 million.

d. Mining Bitcoins

Preview:
- What are the requirements for mining bitcoin?
- The importance of Bitcoin wallet

"When I first heard about Bitcoin, I thought it was impossible. How can you have a purely digital currency? Can't I copy your hard drive and have your bitcoins? I didn't understand how that could be done, and then I looked into it, and it was brilliant."

                                           -Jeff Garzik

Transactions in the bitcoin network are secured because the entire network validates every transaction before it is recorded. Validating transactions require computational power which is provided by nodes in the network using miners. Evey bitcoin transaction has a unique hash function associated with it. The miners should find that uniques hash function whoever finds that first gets rewarded. After the hash of the transaction is discovered it is added to the new block; then this block is verified by everyone in the network to confirm that there is no issue of double spending. The miners are rewarded for contributing computational power, in the form of bitcoins. Bitcoins are thus backed by the computational power used in the network.

In earlier days of bitcoin, it was possible to mine bitcoins using traditional CPUs or GPUs. With increasing demand for higher and efficient processing power, the Chinese tech giants released processors which are far more powerful and efficient for mining bitcoins. The computational power is measured regarding hash rates. The Bitcoin ASICs like Antminer S9 can provide more than 100 times the processing power of a standard processor.

You can consider options like Bitcoin mining through cloud which can primarily contract to entitle you to a particular amount of hash rate in the cloud network. Antminer S9 is the best mining hardware available in the market which claims to provide 13.5Th/s; this is an 8-pound machine which costs around $2000.Once you got your hardware, next step is to sync the hardware with a bitcoin mining software which can be command line software like CGminer or you can use the ones with a proper user interface like EasyMiner.

You are all set to mine now, but you are not set to mine efficiently. Imagine the mining competition you are facing; there are people with thousands of processors invested in mining bitcoins. The reward you can get from mining is always proportional to the hashrate contribution of the node, so you can imagine what your chances against these giants are. There is a solution to this, which is to join a mining pool, which is a group of miners acting as a node and thus increases the possibility if getting rewarded. Your contribution to the pool will split the rewards. The terms and conditions differ from pool to pool, be completely aware of the terms before you start mining in a pool. Slush Pool is an excellent option to join in as a beginner.

The most important rule when you are dealing with bitcoins is to store bitcoins in a suitable wallet. Centralized authorities like bitcoin exchanges and online bitcoin wallets face the risk of being hacked, and it is unwise to store bitcoins in them for an extended period. As soon as you are in possession of bitcoins, move them to a much safer option like hardware wallet. Ledger Nano S is an excellent option for hardware wallet which comes along with a screen where you can check the balance, and you can confirm transactions using a key.

Why should you stay updated with bitcoin-related news?
There is a lot of research and development happening to increase the performance of the processors. If you are missing out on some new resource, you will be missing out on your chance of better investment.
The bitcoin community is getting more centralized with the companies like bitmain becoming the primary source of producing processing hardware for mining bitcoins because of this very reason the community prefers to favor traditional processor based mining; this will ensure that the community will be more decentralized thus ensuring more balance.

Summary:

- Mining requires massive computational power.
- Computational power is measured regarding hash rates.
- Miners use ASIC chips which are manufactured explicitly for mining purposes, Antminer being one of the major vendors in this segment.

7. ETHEREUM

a. Introduction to Ethereum

Preview:
- Comparison between Ethereum and Bitcoin
- Origin of Ethereum
- Basics of Ethereum

After understanding the basics of Bitcoin, several questions pop into our head. Can we get better transaction speed than what Bitcoin currently offers? Is the application of cryptocurrencies like Bitcoin limited to peer to peer transactions?

Yes, the transactions can be improved, and there are broader applications of this technology. Bitcoin played a significant role in introducing the concept of blockchain to the world. The Ethereum introduced the scope of blockchain and the broad spectrum of blockchain's application.

What makes Ethereum so different compared to Bitcoin? And What does it offer which Bitcoin doesn't?

The primary purpose of Bitcoin and Ethereum are entirely different. As we learned earlier, Bitcoin was created as a means of peer to peer electronic cash transactions by eliminating the requirement of a third party. What about Ethereum? It can be seen as an enabler of blockchain technology. Ethereum can be seen as an operating system for constructing and using blockchain based decentralized applications. This feature is enabled by using Smart contracts, which in simple words can be described as condition-based transactions. If you set a smart contract with the underlying condition "Transfer $1000 to John on December 21". On December 21st, $1000 will be transferred to John no matter what. This is the simplest example of how smart contracts. The far superior version can be something like a decentralized online marketplace(analogous to Amazon, which is centralized).

Vitalik Buterin proposed the concept of ethereum in 2013, the Russian-Canadian programmer. Vitalik Buterin is also the co-founder of the Bitcoin Magazine. 11.9 Million Ether were pre-mined for crowdfunding the project in 2014. This crowd sale raised about $14 Million.

The cryptocurrency used in Ethereum network is called "Ether" which is denoted by ETH, as you have seen in the cryptocurrency exchanges. Ether is used for paying transaction fees and to enable computations in the network. Ether is used to buy gas which is the basic unit of computation in Ethereum network. The reason why Ether itself is not used as the computational unit is the volatile nature of the currency.

What is the fundamental difference between Ethereum and Bitcoin?
Now we know that Ethereum provides a broader range of applications as compared to Bitcoin, what about the transaction speed? Ethereum blocks are processed in about 15 seconds as compared to 10 minutes block time of Bitcoin, which is a high plus point for using Ethereum for transactions and also transaction fee is decidedly less compared to that of Bitcoin. Ethereum uses Proof of Work like Bitcoin does but Ethereum's Proof of Work is entirely different and is based on the algorithm Ethash, which doesn't give ASIC mining unfair advantage over traditional processors. We know that the value of Bitcoin halves every 4 years, this doesn't happen with Ether unless there are any hard forks.

Before Ethereum was released, Blockchain applications were meant to do only one thing which was to enable peer to peer cash transactions. The options available for the developers was to either re-engineer bitcoin source code to enable more application or create a new blockchain based platform from scratch, Ethereum was a result of choosing the latter.

Creating a new blockchain network every new application that has to be created is a very time consuming and inefficient process, this issue was solved by Ethereum by introducing the Ethereum Virtual Machine which enables testing of applications without the requirement of creating an entirely new blockchain. Using this platform, many such smart contracts can run simultaneously without hindering the original blockchain network. Ethereum Virtual Network is essentially a sandbox which is isolated from everything else(Ethereum network or the node owner's system). EVM is implemented in every Ethereum node, and it acts as a consensus mechanism for the Ethereum network.

Can you code smart contracts with your previously learned coding skills?
The smart contracts are mainly coded using the programming language known as Solidity which can be easily adapted to if you have a base in C and Javascript. There is a language known as Viper based on Python which is being developed, which will soon be replacing the Solidity.

Smart contracts are not directly deployed in the network due to the constant and immutable nature of the network. The ledger can't be changed due to the action of any bugged smart contracts, to prevent such issues they are first perfected in the virtual machine environment.

The hack on DAO was an incident where an anonymous hacker exploited a bug on the smart contract.

Summary:
- Ethereum enables a wide range of blockchain applications.
- Ethereum Virtual Machine provides an environment for testing smart contracts
- You can get started with coding Ethereum by learning Solidity

b. Smart contracts

Preview
- Using smart contracts in your real estate
- Defining a Smart contract
- Smart contracts in Insurance
- Smart contracts in copyright

Imagine, after investing in cryptocurrencies you made a hell lot of money, after giving a lot of thought where to invest this money, you finally decide to try your luck in real estate. You have always fancied owning a bunch of houses with pools in New Jersey. Now you invest in 5 of those, one for yourself and the others for renting out. After a couple of weeks, you manage to get tenants for the others properties, and according to the agreement, rent will be paid on the 1st day of every month. After a couple of months, you realize that it is a great ordeal to collect rent from these people and thus to waste much of your precious time and to hinder your peace of mind. Each time you try to collect rent, they will be trying to extend the date by telling excuses which you find very hard to digest and you get a gut feeling they have the money at the time of your rent reminder visits.

When you are thinking and searching hard to find a solution to this problem, you come across the idea of smart contracts. When you understand the idea that smart contracts are transactions based on conditions, you get excited suddenly because of a Eureka moment by realizing " It can be implemented for collecting rents regularly."

Now, you are buying a set of digital locks, which can be opened by entering digital keys which are programmed to change every month. Now with the help of your expert coder friend Jack, you are building smart contracts with the goal "Every month when the tenants pay the rent, they will receive the new digital key for accessing their respective houses."

Now all you gotta do is sit back and relax, so that money pours in automatically every week. If the tenants are not paying the rent, they won't be able to access the houses. They will leave with no other option than to pay the rent in time.

From your real estate experience, what is the best definition of a smart contract you can think of?

Smart contracts can be used for purchasing and exchanges of property or anything which possess value without any involvement of a middleman. The smart contracts can be visualized as a vending machine, which delivers the product with some value when feeding our money in it.

If the case of real estate is considered as a vending machine, the product that comes out is the digital key, which is renewed every month. You can even set the penalties which get enforced automatically as the due date passes by. The fact that these penalties get automatically enforced will increase the chances of the tenants paying on time.

Do you have any experience with insurance policies?

They are very laggy processes which can sometimes take months to get processed. These policies require a great deal of manual activity which makes it cumbersome and complicated. This other lengthy processes also add up to the cost due to the effort put into, which is to be paid by the customer. What is the solution, if you want to take the complexity out of the insurance scene. Yes, like the case of real estate program everything in smart contracts so that when an event occurs, the program will be automatically triggered to release funds. Most of the natural calamities like an earthquake can be measured, and the damage can be predicted to an extent. By setting the smart contract in such a way that it gets activated when an earthquake above a particular Richter scale hits, a lot of human labor can be avoided. This can significantly increase the transparency of the processes when the ledger is distributed throughout the network, and the trigger parameters are predefined. Immutable and Distributed nature of the smart contracts will also ensure that there is no discrepancy and manipulation in the data after the event has occurred.

What do you think about enforcing a copyright situation? How difficult is to enforce legality each time copyright is breached?

When you are creating some original content which is copyrighted, each time sometimes reproduces the content; you deserve your share of royalty to be paid. Sometimes the rights are distributed among a group of people with different weighting. A smart contract can be established which can be used to preset real owners of the content such that each time it is reproduced, the smart contract will be activated which will force them to either pay the royalty or to revoke the content. This process will ensure the timely payment of the royalties and the reproduced content will not stay under their name for a long time. This being a distributed record, the other member of this network can confirm that the content is indeed reproduced. Imagine the number of legal procedures and the magnitude of expenses which will go behind this process if we are traditionally handling this.

Smart contracts are still in its growing stage. Bugs in smart contracts can cause a lot of issues. If you are a tenant, you left your house a couple of weeks before for a trek. The place where you are trekking has no network coverage to enable payment. The bank account which was linked to your smart contract ran out of balance due to some unforeseen circumstance. The smart contract was designed in such a way that if rent is not paid in time, the house will be going public for the next tenant. This is a helpless situation for you because the smart contract is enforced, no matter what.

Summary:
- You can enable smart contracts to automate transactions like paying rents
- Smart contracts are condition based transactions
- Smart contracts can be used to trigger insurance claims instantaneously
- Copyright privileges like collecting royalty and taking down infringing materials can be enforced using smart contracts.

c. The hack on DAO

Preview
- DAO as a Venture capital alternative
- Working of DAO
- The hack

If you have a promising startup and you require investments, you can trade a part of your company's equity in exchange for investment which is known as Venture capital fund. These kind of investments are mostly the kind which can be described as high-risk/High-reward. The investments usually depend on characteristics of the company like its size and stage of its research and development. These funds are mostly early-stage investment or a means of expansion.

What if your startup is based on decentralized applications? This is where DAO or Decentralized Autonomous Organization comes in, DAO acts like a venture capital fund for startups based on the blockchain. The main difference when compared to other typical venture funds is the absence of a centralized entity. This lack of centralization will give more control to the investors. There is no management hierarchy involved in DAO like the typical venture capitals. Since DAO is a decentralized worldwide phenomenon, it is far beyond the government regulations. In May 2016, DAO was announced, just like the first block is known as genesis block, this was known as Genesis DAO. The crowd sale was a huge success by raising around $150 Million, and this was the most successful crowd sale in the history of all of the crowd sales.

How does DAO work?
DAO was created to be a fund dispersing medium for startups and projects eliminating the requirement of the board of directors. The "Contractors" submitted proposals for projects which are validated by the "Curators." The curators whitelist the project if they find it to be legitimate. The DAO token holders are essentially the stakeholders who can vote on these projects once they are whitelisted. If the voters approve the project, the funds will be released to the contractors based on their proposed timeline. The funds won't be released as a lump sum; it will be distributed throughout the timeline in a well-balanced manner. The DAO also enabled a mechanism to exit from an investment, which would return the investment to the investors. This process known as splitting is like Ctrl+Z in the DAO funding process.

DAO sounds like a very promising initiative, but it had its drawbacks and shortcomings. The most notable one was the issue of funds being drained from the DAO network. We just learned the mechanism using which the investors can take their tokens back using the Splitting functionality. There was a significant bug related to this feature which allowed the hackers to withdraw funds multiple times. The hackers kept requesting the invested tokens before the network could update the balance details, this caused the network to keep dispensing the money. This bug was known as recursive call exploit. The hackers drained around $70 Million worth of Ethereum within the first few hours of exploiting this recursive call bug. This can't be seen as the shortcoming of the Ethereum network, these bugs and the many others associated with DAO are due to the drawbacks in the DAO code.

All the mess around the DAO initiative did split up the Ethereum community with one group supporting the returning of DAO tokens and the others supporting. Those who supported the return of DAO tokens were arguing that the hacker who drained all the token shouldn't be benefiting from the hack and this is only possible by returning the DAO tokens to the investors. This reversal will stay as an example and prevent future exploits of such scales. The return of tokens will enforce the fact that any illicit activity can be controlled and when a hacker is possessing such a significant portion of DAO tokens, it will harm the network. Those who did not support the decision to return the token were strongly believing that the underlying ideology of blockchain- immutability, which should be protected no matter what and the process of returning the token will drastically decrease the trust on Ethereum in future projects.

The result was a hard fork with about 90 percent of the community supporting the return of DAO tokens, and thus the split up community created the new Ethereum Classic.

Summary
- DAO is the venture capital fund for startups and projects based on the blockchain.
- The Curators in the network decide the legitimacy of the projects.
- A bug in the DAO network allowed hackers to withdraw currency multiple times from the network.

d. The Turing complete Ethereum network

Preview
- The significance of Turing completeness
- Importance of Gas in Ethereum Network

When a system or a program can compute anything when it is provided with enough resources, then is said to be Turing complete. Ethereum executes a transaction till the gas allocated to that particular transaction gets exhausted. This happens following the Post's Theorem which states that it is more efficient to provide information and to predict the outcomes before the event happens. In Ethereum transactions, the amount of gas which is to be allocated to each transaction is mentioned beforehand. The Gas is the measure of computation in the ethereum network, so when gas is mentioned in the transactions, it is predicting the outcome of the transaction which will increase its efficiency as mentioned in the post theorem.

What happens when you try to execute an infinite loop using a smart contract in Ethereum network?
The program would not run forever if you expected that to happen. As mentioned earlier, gas is required for enabling computation in Ethereum network. The amount of gas associated with every transaction is mentioned beforehand. This will ensure that if the program functions unexpectedly, it will get terminated once the specified gas is consumed. The program will be killed before it goes haywire.

Gas maintains a balance in the Ethereum network, by checking the gas associated with a transaction we can decide whether the transaction fee associated with the transaction is justified. Gas is like a prepaid computational meter. As we know the reward system, where the transaction fee motivates the miners to stay active in the community and provide computational power to the network.

For a transaction to process, it is a necessity that the amount of gas should be specified along with the transaction. The amount of gas required to complete the process will entirely depend on the complexity of the process or the computational power required to complete the process. If the process is completed far before the gas is exhausted, the remaining gas will be refunded to the initiator of the transaction. If the gas runs out before the process is completed, the miners will stop mining the transaction.

If you set the gas limit to be very low in a transaction, you will still have to pay the transaction fee irrespective of whether the transaction is completed or not. This is because setting a low gas value is a fault on your part which shouldn't compromise the activity of the miners. This doesn't mean that you can keep the gas value high and expect it to be returned because miners will not accept transactions tagged with high gas values. The miners avoid such transactions because there is a limit on how much gas the miners can be exposed to in a block which is known as block gas limit.

When we are considering both gas and the associated transaction fee. The miners will the transactions with just enough amount of gas and the transaction fees which is equal to or above the value which justifies the amount of computational power involved.

Summary
- The gas serves as the computational fuel in the Ethereum network.
- Gas ensures that the fees for the corresponding transaction are justified.
- The gas feature ensures that any infinity loops will be eliminated.

8. ALTCOINS

Preview:
- Introduction to coins other than Bitcoin - Litecoin, Ripple, Bitcoin cash, EOS, Stellar and Cardano

Every cryptocurrency other than bitcoin is known as altcoin since we already dealt with Ethereum in detail let's explore other cryptocurrencies.

If Bitcoin is the gold in the crypto world, then Litecoin is considered to be silver. The Bitcoin code can be tweaked to create new coins, Litecoin was created by modifying bitcoins source code. Two of the issues related to bitcoin is the transaction time and the mining algorithm which favors only people with substantial resources. With the better transaction time of 2.5 minutes, Litecoin can process a large number of transactions. As opposed to the SHA-256 algorithm of bitcoin, Litecoin uses Scrypt which requires more memory to operate. This feature will ensure that the mining giants won't hold the monopoly of the Litecoin network. Anyone with traditional processors preferably GPUs(which can process more than CPUs) can mine Litecoins. This will ensure that the network is decentralized because if mining giants dominate the network, the companies producing the mining equipment will start to have a monopoly over the network.

Ripple was the cryptocurrency introduced to revolutionize the financial system. Unlike bitcoin ripple is not concerned with peer to peer electronic cash transactions but it is used by facilities like banks to enable better interbank transactions. The ripple transactions cost negligible fee. The ripple is more of a common currency enabling cross-border transactions. Unlike bitcoin ripple is more of a centralized because the Ripple labs hold most of the XRP tokens that were pre-mined. Ripple can act as a bridge from centralization to decentralization since it is a combination of both. It is relatively easier to digest the application of ripple as compare to other decentralized networks.

We know what one of the issues with bitcoin was its low transaction volume, the bitcoin community was split on deciding whether the block size of the network should be increased to solve the issue or not, since consensus was not reached with regard to this decision Bitcoin cash was created with a block size of 8 MB as opposed to 1 MB block size of bitcoin. As we know that the mining difficulty depends on the computational power contribution to the network, as there are fewer participants in bitcoin cash network, it will be relatively easier to mine than bitcoin.

EOS operates more like Ethereum by providing an environment for developing decentralized applications. This network can enable up to a million transactions per second using negligible latency and transaction fees. EOS coin is the cryptocurrency used in EOS network; it will be required for running applications in the network. EOS is more user-friendly than Ethereum network because it allows users to code applications in pre-existing coding languages, unlike Ethereum which demands the knowledge specific language Solidity which could be switched to Viper in future. The consensus mechanism used in the network is delegated proof of stake. EOS lets you create a hierarchical system with various levels of permission accessing. EOS also enables server hosting and cloud storage. The decentralized commercial applications would require a lot of communication channels, EOS provides parallel execution and asynchronous communication to make this happen.

Stellar can enable transactions in just a matter of seconds. XLM or lumen is the cryptocurrency used in the Stellar network. Spamming in the network is avoided by incurring negligible transaction fees as low as 0.00001 lumens for each transaction and also every user in the network must have a minimum of 20 XLM to operate in the network. People with no access to bank facilities can significantly benefit from the affordable transactions, transparency, and ease of use of the Stellar network.

Another smart contract platform is Cardano which solves the scalability issue by following a layered structure. Cardano was designed by Charles Hoskinson who is one of the co-founders of Ethereum.

Summary
- You can check out cryptocurrencies based on market capitalization, which is the total value of the cryptocurrency in the market which can be calculated by multiplying the total number of coins by the present value of the coin. By referring to this list, you will get a better idea of which cryptocurrency to learn more about and root for.

IOT GUIDE

1. Introduction

Preview:
- Definition of IoT
- How did it originate?
- How is it useful?

"The Internet of Things is not a concept; it is a network, the true technology-enabled Network of all networks."

– Edewede Oriwoh

Introduction to IOT

We are living in an era where everything around us is getting connected to each other. Gone are the days were machines required human intervention to communicate with one another. Today with the explosion of data and advancement in technologies the capabilities of machines are being extended to the next level. IoT is an enabler for these kinds of developments at the root level, and it is driving the next wave of digital transformation. Today, technology is roiling the landscape of businesses, and IoT has a commendable role to play in it.

Businesses embracing technology have disrupted the market in a matter of time. For instance, companies like Uber have disrupted their related industry overnight without focusing on traditional business ideas. Uber innovated by introducing a new business model. They connected the drivers with the passengers and Boom! The taxi industry was disrupted. The tremendous impact was caused by this disruption, and the paradigm shift came to be known as uberization, a great way of disrupting the traditional businesses by today's idea economy.

The Origin of IoT

The process of computing started with calculators where numbers were the only required inputs, with the increasing data processing capacity of the computers the sources of information grew along with it. During 1880, Herman Hollerith found a method for collecting and processing US census data using punched cards. Recording census is a tedious process, the number of attributes associated with being the reason. These punched cards made the process easier and more arranged to catch up with the growing population. This punched card is one of the first use of data processing, the origin of the internet of things can be dated back to this event.

The first smart device was designed in 1982 by Carnegie-Mellon students; it was a vending machine connected to the internet. This allowed the students to check the status of the machine like knowing its current stock and temperature. There were plenty of data available before the internet of things which were not ready to be used and optimized. With the entry of sensors into the scene, the machines started communicating and conveying the data among themselves. Now the system could be optimized using this data, and the value of those data was understood. How can the product be improved to better suit the customer needs? How can the data be used to improve the efficiency of processes to cut cost? Everything was started to get the answer. The interesting fact that these sensors were available at a lower loss and its size was getting reduced with more research that came along with the demand. The tools which were used for analyzing the online traffic like NoSQL and Hadoop was found out to be suitable for processing data from smart devices.

What is IOT?

So in today's world where everything is interdependent, the availability of data and the ability to process it in real time is highly significant. Further, it is necessary to share this information between different levels of the networks without any latency. The delay between instruction for data transfer and the actual process of data transfer is known as latency. The development in Internet technologies has helped in connecting devices with each other, enabling the communication between them in real time.

IoT is all about using embedded electronics, software, sensors, actuators and data communication technologies to build a network of devices, appliances, and other items, referred to as "Things" in IoT. The "Things" collects data as well as exchange them over the network. There are three steps involved in an IoT application,

- Capturing data from the object (for example, simple location data or more complex information),
- Aggregating that information across a data network, and
- Acting on that information - taking immediate action or collecting data over time to design process improvements.

IoT is completely transforming the way traditional businesses worked. Let's see how IOT does that:

Invent new business models

With the transformative opportunities of IoT, companies are reinventing different business models which can increase their bottom line. As discussed at the beginning of the chapter, traditional businesses have to adapt with the emerging technologies to eliminate the potential threat of extinction. Uber revolutionized the taxi industry by utilizing the benefits of the Internet of things and shared real-time data of the vehicles to its customers. Today they have expanded all over the world, and many companies have replicated this business model since then. This has made a significant impact on traditional taxi companies. So, Today, the line of businesses have to closely interact with IT teams to adopt new business strategies and stay relevant in the market.

Improve operational efficiency

IoT systems can be of great use to improve the operational efficiency. Companies today are using sensors to gather a large amount of data, and this data can be utilized to streamline and automate business processes. IoT systems are used in many devices to gather the health status of them in real time which can be monitored in real time, thereby decreasing the downtime. Rolls Royce has built an Engine Health Management System in its aircraft engines which gathers terabytes of data over the course of a flight. This data is then analyzed and utilized to improve fuel efficiency, identify maintenance issues and reduce other operational costs. In industries like manufacturing, IoT can improve the process flow by integrating the production systems. Further, the data that is collected can be utilized to get insights which can further fine-tune future operations.

Drive better customer engagement

IoT also helps businesses to interact with their customers efficiently. Companies are trying to engage with their customers and built the relation ahead. Most of the companies focused on selling the product, and the kind of business they were doing was transactional. However, today with the advancements in technology and the availability of real-time data companies can offer many services to its customers.

The famous truck manufacturer John Deere is an excellent example of this. Their business was the only transaction based. The only focus was about the tractor. However, today that tractor is attached with sensors to offer farmers services like reporting on average fuel consumption and engine performance. This gives them the ability to keep continuous engagement with the customer by offering better services. Imagine what internet of things could do to your industry?

Summary:

- IoT refers to physical devices and appliances which have embedded electronics, software, sensors and actuators and are connected to each other with internet which enables them to collect and exchange data.
- IoT is a disruptive technology and is transforming businesses drastically enabling them to capture new markets.
- IoT helps businesses to improve their operational efficiency, drive better customer engagement and event to invent new business models.

2. Types of IOT

Preview
- Different types of IoT
- Definition of Consumer IoT and use cases
- Definition of Industrial IoT

"Security by design is a mandatory prerequisite to securing the IoT macrocosm; the Dyn attack was just a practice run."

– James Scott, Sr. Fellow, Institute for Critical Infrastructure Technology

Based on the target market, IoT is broadly divided into two: Consumer IoT and Industrial IOT.

Consumer IoT, as the name, suggests the target make life more convenient for individuals. It comprises connected devices which target the consumer market. With the internet becoming a basic necessity and with increasing connectivity, the growth of IoT in the consumer segment is tremendous.

They are of many types starting from interactive toothbrushes to smart door locks, but the most popular ones being activity trackers and smartwatches. These smart devices intend to make our life easier, safer, healthier and smarter.

However, the implementation of IoT in the consumer segment has not yet become as successful as in the industrial sector. The following are some of the consumer IoT use cases:

Smart Wearable: Wearable technology refers to electronic devices that can be worn on the body as implants or accessories. These devices use sensors and embedded electronics to generate data and then enable the exchange of data between the network and device through the internet.

Wearable technology has a variety of applications which grows as the field itself expands. It appears prominently in consumer electronics with the popularization of the smartwatch and activity tracker. Apart from commercial uses, wearable technology is being incorporated into navigation systems, advanced textiles, and healthcare.

Smart Homes: This is one of the trending and most popular application of consumer IoT which makes use of internet-connected devices to enable remote monitoring and management of home appliances and systems. It is also referred to as home automation and aims at increasing security, energy efficiency, and comfort.

Some of the examples of smart home technologies include Smart irrigation, smart locks, smart lights, smart thermostats, smart TV and smart security.

The industrial internet of things, or IIoT, enhances industries and businesses by converging information technology and Operational Technology, thereby paving the way for the revolution of the industrial sector. Thus, It is also referred to as the "industrial internet" or Industry 4.0. The primary objective of IIoT is to increase operational efficiency and to transform business processes in various industries. IIoT serves a wide range of industries like manufacturing, logistics, automotive, utilities, cities, agriculture, healthcare, retail, and IT and the market is estimated to be approximately 123.89 billion USD by 2021.

The Industrial Internet Consortium defines IIoT as "machines, computers, and people enabling intelligent industrial operations using advanced data analytics for transformational business outcomes."

The driving factor of Industrial IoT includes intelligent machines or smart machines and advanced data analytics. Sensors and software applications are integrated to machines and facilities to gather real-time data which is analyzed using advanced data analytic tools. This data can be utilized by companies to tackle inefficiencies.

The manufacturing sector is the largest market for IIoT, IIoT, and companies today are widely adopting it in their business strategy. There are many benefits of integrating IIoT into the production equipment. By introducing automation and more flexible production techniques manufacturers can boost their productivity.

Today, IIoT is transforming companies, and it opens up a new era of economic growth and competitiveness. By incorporating AI, analytics and big data, the data generated from IOT can be harnessed and used for machine to machine communication and automation. Companies are creating new hybrid business models and, using intelligent technologies to fuel innovation and workforce transformation. IIoT has two ways of being realized:

- Through the manufacturing process of a product
- Through the product itself

Below listed are 15 possible uses of IIoT, defined by Industrial Internet Consortium:

- Smart factory warehousing applications
- Predictive and remote maintenance.
- Freight, goods and transportation monitoring.
- Connected logistics.
- Smart metering and smart grid.
- Smart city applications.

- Smart farming and livestock monitoring.
- Industrial security systems
- Energy consumption optimization
- Industrial heating, ventilation, and air conditioning
- Manufacturing equipment monitoring.
- Asset tracking and smart logistics.
- Ozone, gas and temperature monitoring in industrial environments.
- Safety and health (conditions) monitoring of workers.
- Asset performance management

Challenges in IIoT

Security: Data is crucial for almost every businesses, and data security is a growing concern today among companies. Weak network layer and security systems can keep devices exposed to threats from external attackers.

Lack of standardization: Industries comprises of different types of machines with different protocols. There is a wide range of designs and standards for everything from transmission protocols to ingestion formats. For example, if a device that sends operational information about the temperature of a boiler isn't made by the same vendor that makes the network or the data ingestion engine, they might not work together.

Integration with legacy technology: Old equipment is not compatible with IoT networks and cannot provide data in the format they require. It will be tough to integrate a power station controller which is 10 years old to a sophisticated new IIoT infrastructure.

Lack of skills: To gain the maximum out of IIoT, it often requires expertise in many areas like network systems, real-time analytics, and data science.

Investment: IIoT implementation would incur a massive investment for industries and traditional businesses since it demands newer machines, hardware, software and skill sets. Even though IIoT is going to cut down on costs, in the long run, businesses will have to incur an upfront capital expenditure.

Summary

- IoT is broadly divided into consumer IoT, and Industrial IoT based on the target market addressed.
- Consumer IoT comprises connected devices targeting the general public and aims at making life more comfortable and convenient.
- Industrial IoT is a broader market and involves IoT applications for the enterprises.

3. Prerequisites

Internet of things is about giving an identity to many objects. We are bringing a significant gap with IOT. The expectation of humans to extract information from the machine is eliminated when it is given an identity and a power to communicate for itself. The objects will communicate with other objects and the first frame where the data is required. However, how can you make this possible? What are the fundamental knowledge requirements for swimming the shores of the internet of things? Can you be aware of the sensors available in the market and be an IOT master? No, you can't be, because for getting a full picture of the internet of things, you should be aware of:

- Sensors
- System on Chips, Microprocessors, and Microcontrollers
- Embedded Systems
- Protocols
- Basic understanding of programming languages like C or Python.

It is not necessary to be an expert of everything mentioned here and it not expected. Awareness can increase your choices, and that is the point of having a basic understanding of these categories. By being aware of all these, you will have an option to specialize in one or more if that catches your interest by any chance.

4. Sensors and Actuators

Preview:

- Definition of Sensors and actuators
- Various types of sensors and its application

"The Internet will disappear. There will be so many IP addresses, so many devices, sensors, things that you are wearing, things that you are interacting with, that you won't even sense it. It will be part of your presence all the time. Imagine you walk into a room, and the room is dynamic. Moreover, with your permission and all of that, you are interacting with the things going on in the room."

-Eric Schmidt

Building Blocks of The "Things "in an IoT network: Sensors and Actuators

Sensors and actuators form the building block of IoT, and they are the fundamental requirement for making devices smart. The smart devices in IoT or referred to as the "things" have embedded sensors and actuators, which can sense physical quantities, convert those measurements into a digital representation, and trigger a physical effect after analyzing this data. Let's briefly understand about different kinds of sensors and actuators.

i. Sensors:

Sensors, as the name indicates are physical devices that use embedded technology to sense and measure practically any measurable variable in the physical world. In short, a collection of various data in IoT devices are done by sensors. There are different types of sensors such as humidity sensors, temperature sensors, pressure sensor, light sensor and many more. These sensors collect the data and send them to a database or a data warehouse for further analysis and take specific actions based on that.

It is essential to keep in mind a few critical factors while choosing the right sensor for your IoT deployment.

Cost: It is essential to consider the cost of the sensors to ensure that the overall solution is genuinely cost-effective.

Accuracy & Precision: It is essential to ensure that the sensors can offer the degree of precision and accuracy required for the specific application. For example, monitoring biological stock in an industrial farm will require more precise temperature sensors than monitoring the temperature in your bedroom.

Measurement Range: Another factor to make sure is that the range applies to for what you want to use it. It would be foolhardy to get a temperature sensor that only works above 40°C to monitor freezers.

Power Consumption: The power consumed by the sensor consumes should be compatible with the system. There are chances that the sensors would be blown out if the circuit has too much current.

Categories of sensors

Active or passive: Active sensors require an external power supply and produce an energy output whereas passive sensors do not require an external power supply.

Invasive or non-invasive: Invasive sensors part of the environment it measures whereas non-invasive sensors are external to the measuring environment.

Absolute or relative: Based on whether they measure on an absolute scale or based on a difference with a fixed or variable reference value sensors can be classified as absolute or relative.

Let's take a look at some of the critical sensors, extensively being used in the IoT world.

Temperature sensors

Temperature sensors are much every day and have been used in a variety of applications like A/C temperature monitoring and control, refrigerator and other similar devices. With the advent of IoT, they have found more room to be present in an even higher number of devices.

Today, they have found their role in manufacturing processes, agriculture and health industry. Certain machines require a specific environment and device temperature to operate during the manufacturing process. Temperature sensors are deployed to ensure that the temperature is always kept within that specific range, thereby optimizing the manufacturing process. Temperature sensors are used in agriculture to measure the temperature of the soil, which is crucial for crop growth. This helps in taking the necessary steps to maximize the output.

Proximity sensor

The automobile industry is rapidly growing with more and more vehicles being added to the road on a daily basis. Further, serious researches are going around self-driving vehicles and making it possible for vehicles to drive you and not the other way around. This can be achieved only by making use of a lot of sensors, one of the most important one being proximity sensors. In combination with all the other sensors on the vehicle, it ensures that the vehicle remains away from any obstacles from all sides of the vehicle.

Another application of proximity sensors can be in smart retail stores. In retail stores they can be used to determine the correlation between the customers and product they are interested in and thereby notifying them about particular product characteristics, discount offers and so on.

Pressure sensor

Pressure sensors are of great use, especially in the manufacturing industry where there are steamers, boilers and other devices that rely on liquid or other forms of pressure. Pressure sensors are used to deploy IoT systems that monitor systems and devices that are pressure propelled. Once a drop in pressure occurs, the device notifies the system administrator about any problems that should be fixed.

Water quality sensor

Water quality sensors are used to measure the chemical concentration of water or to measure substantial deposits in it. They are gaining popularity since water is highly precious and the demand for pure drinking water is increasing day by day. Further, there are industrial use cases where pure water is required. Therefore, water quality sensors are playing an essential role in monitoring the water quality for different purposes.

Chemical sensor

Chemical sensors are devices used to convert chemical information into analytical signals. Different kinds of chemical information constitute of factors like the presence of a particular element, any chemical reaction, change in concentration of a particular chemical and so on. They are used in various industries, and the primary objective is to indicate changes in liquid or air chemical changes because of which they have applications in smart cities with a large population to track changes and protect the population.

Gas sensor

Gas sensors are a type of chemical sensor and are correctly used to monitor changes of the air quality and detect the presence of various gases.

Smoke sensor

Smoke sensors are an integral part of high rise buildings and official spaces along with fire security alarm and have been used for an extended period. Today they are used to deploy intelligent fire protection systems by integrating with IoT systems. Whenever the quantity of smoke exceeds a particular limit, the information can be passed on to property owners, safety guards, and firefighters through a backend management platform.

Besides the accommodation industry, these sensors are used in manufacturing, where there is a high risk of fires. This serves to protect people working in dangerous environments, as the whole system is much more effective in comparison to the older ones.

IR sensors

IR sensors are electronic devices which can sense specific characteristics around its environment by the emission and detection of infrared rays. They are used to measure heat emitted by a body, detect motion and detect shape. They have found its application in the health industry and are used to make monitoring of blood flow and blood pressure simple. They are even used in various smart devices such as smartwatches and smartphones.IR sensors are also used in implementing home security, and to detect chemical and heat leaks.

Level sensors

These sensors are used to measure the level of liquids or fluids or fluid-like solids which can have a free upper surface like powders, slurries, and grains. These sensors are useful whenever gravity makes the substances to occupy a horizontal position in the container. Recycling, juice and alcohol industries greatly depend on level sensors to measure liquids in different sections of the industry, so that the production can be optimized and thus increasing production efficiency.

Light Sensor

Light sensors are used to detect light. For example, a robot might make use of a light Sensor to detect how dark or bright it is. There is a range of different types of light sensors like photocells, Photoresistors, Photodiodes, and 'Phototransistors.

Light sensors have many applications. They are used in safety or security devices like a garage door opener or a burglary alarm. Modern electronic devices such as smartphones televisions, laptops including TV's, computers etc. use a different type of light sensors to automatically control the brightness of a screen in situations where light intensity is high or low.

Image sensors

Image sensing has got large applications in various industries like healthcare, transportation, and many more. Image sensors are used to detect the details of an image, collect the data, interpret them and take necessary actions making use of this data.

With much research going around driverless cars, image sensing would be finding many applications in the automobile industry as well. With these sensors, the system can recognize signs, obstacles and many other things that a driver would generally notice on the road. Omron and Envisage are two major companies who are a pioneer in implementing image sensing technologies for IoT.

Motion detection sensors

Motion detection plays a vital role in the security industry. Businesses utilize these sensors in areas where no movement should be detected at all times, and it is easy to notice anybody's presence with these sensors installed.

On the other hand, these sensors can also decipher different types of movements, making them useful in some industries where a customer can communicate with the system by waving a hand or by performing a similar action. For example, someone can wave to a sensor in the retail store to request assistance with making the right purchase decision.

Even though their primary use is correlated with the security industry, as the technology advances, the number of possible applications of these sensors is only going to grow.

Accelerometer sensors

Accelerometers measure the proper acceleration of the object to which it is attached. Proper acceleration is the object's acceleration when we are considering its rest frame of reference. These sensors are used in huge volumes, in millions of devices like smartphones. The accelerometer can also be a safety mechanism which can trigger alert when the object is displaced. The user interface can be controlled using the accelerometer. The motion input can also be used for playing motion-based games.

Gyroscope sensors

The objective of the gyroscope is to detect any rotation or twist. Gyroscopes come along with accelerometers to provide better feedback to the system. This sensor can be used by athletes to map their movements so that they can improve the efficiency of their activities. The typical example can be observed when the phone screen self-adjusts to landscape or portrait according to its orientation.

Humidity sensors

Humidity sensors detect humidity or the amount of water vapor in the air or a mass. There are many manufacturing processes which require perfect working conditions and specific humidity levels. By utilizing humidity sensors, you can ensure that the whole process runs smoothly, and when there is any sudden change, action can be taken immediately, as sensors detect the change almost instantaneously.

ii. Actuators

While sensors sense physical variables, actuators, on the other hand, triggers a physical action by an input signal. The data measured by sensors are processed, and this processed output will become the input for the actuator. There are different types of actuators from the type of motion, output power and area of application. For example, the actuator can be a simple valve, which can open or close based on the input received. The actuator requires a power source and signal. Upon receiving the signal, the actuator converts it to mechanical energy.

Fluids are almost impossible to compress which makes them suitable for mechanical operation. The motion can be linear, rotary or oscillatory but there's a limitation for the maximum acceleration which can be achieved using hydraulic actuators.

Pneumatic actuators facilitate linear and rotational motion by using the energy of compressed air. This component is used in controlling engines since they can respond very fast on command. Compared to other actuators, pneumatic actuators are more affordable, safer and powerful.

Electric actuators convert electrical energy into mechanical torque, and there are no fuels involved directly in its working which makes them the cleanest form of actuators.

Mechanical actuators convert one form of mechanical action to another, the best example is the action of rack and pinion which converts rotational motion into linear motion.

The attributes like force, speed and acceleration can be used to increase the performance of the actuators. The force can be static when the actuator is at rest or dynamic when the actuator is moving. The speed of the actuators decreases with the load. The trend of how the speed is affected by the load can be used to optimize the actuator.

Summary:
- Sensors are physical devices that can sense and measure variables in the environment.
- Temperature sensor, pressure sensor, humidity sensor, light sensor are different kinds of sensors.
- Actuators trigger a physical action from an external input signal.

5. System on Chip

Preview:
- What is System on the chip?
- Methods of fabrication

"The global industrial sector is poised to undergo a fundamental structural change akin to the industrial revolution as we usher in the Internet of Things. Equipment is becoming more digitized and more connected, establishing networks between machines, humans, and the Internet, leading to the creation of new ecosystems that enable higher productivity, better energy efficiency, and higher profitability. While we are still in the nascent stages of adoption, we believe the Internet of Things opportunity for Industrials could amount to $2 trillion by 2020. The Internet of Things has the potential to impact everything from new product opportunities, to shop floor optimization, to factory worker efficiency gains that will power top-line and bottom-line gains."

-Goldman Sachs

We have our embedded systems in place, but how are they integrated - that happens using the System on a Chip abbreviated as SoC. This same thing is used for integrating components in computers and other electronic devices. Soc will make the system works as if they are powered by a motherboard. This single integrated circuit will be exhibiting high performance and a reduction in power consumption.

Let us see what constitutes SoC:
- Microprocessor, microcontrollers or DSP
- Memory blocks
- Clock signal generators
- Peripherals
- External interface
- Analog interface
- Power managers

The working of SoC is controlled by protocols like the ones driving the standard USB. The software is integrated using software development environments.

Fabrication of Soc is done by 3 different methods

- Standard cell ASIC
- Field Programmable array
- Customized ASIC

The System on Chips reduces the power consumption considerably when compared to the complicated systems they are replacing. They offer high reliability, and the manufacturing cost is low.

Summary:

- System on Chip is an integrated circuit that integrates all the components of a computer or an electronic system.
- Considerably reduces the amount of power consumed.

6. Embedded system

Preview:

- What is an embedded system?
- How are they used in IoT?

"We need to get smarter about hardware and software innovation to get the most value from the emerging Internet of Things."

– Henry Samueli

Internet of things is the ecosystem where the embedded systems communicate with each other. IOT is like the superset of an embedded system.

Embedded systems are not like computer, and they are expected to perform a specific task. In some cases, they are required to act at high performance especially when safety is concerned. Embedded systems are very minimalistic to cut the costs, and this is made possible since they are only expected to perform a specific task which helps the researchers to optimize the device better. They are not stand-alone devices and are a part of the bigger system. About 98 percent of the microprocessors in the entire world are manufactured as an embedded system, well that should have given you the degree of influence. The programming within the embedded system is referred to as firmware which runs in limited computational resources and stored in read-only memory.

Look around, and you can easily find the applications of embedded systems to increase the efficiency and flexibility of various devices. You can find embedded systems in your mobile phones, DSLR, video game consoles, washing machines, MP3 players. You must have heard of the Anti- Braking System(ABS), how do you think that's building? They play a significant role in improving safety features of transportation systems ranging from the inertial guidance systems in flights to automatic four-wheel drive.

Embedded system can act independently from the system which gives them the ability to function when there are communication outages and loss of electrical systems, and this makes them a great candidate for safety features. The systems can be designed to tolerate high temperature in case of fire outbreaks. Let's take the case of the health sector; the embedded system is used in monitoring vitals, amplification of teeth sounds and imaging using noninvasive methods like CT and MRI scans.

As mentioned earlier, embedded systems are a part of a bigger system, so they require minimal or no user interface. In simple embedded devices, it is mostly with buttons and LEDs powering a simple menu. In more complex devices touchscreens are provided for handling the embedded system. The interface should not be necessary within the device, and it can also be linked to a PC, just like we configure our router while setting it up.

The embedded systems are expected to function for ages without any errors, and they recover if some error occurs by chance.

Summary:

- An embedded system is a computer system dedicated to performing a specific function.

7. MEMS

Preview:

- Definition of MEMS?
- Manufacturing of MEMS
- Applications of MEMS

How does inkjet printers operate? How are airbags deployed in modern cars?

Microelectromechanical Systems (MEMS) is the answer. They are little devices which can come with moving parts. The components of mems come in the range of 1 to 100 micrometers

The manufacture of memes follows the same basis as that of semiconductors like deposition, patterning, and etching. Silicon, polymers, metals, and ceramics are used to fabricate MEMS. Silicon is affordable, and it can be incorporated with electronic features which makes it a viable material choice. Polymers can be manufactured in high volumes, and they can cater to a vast range of requirements. Metals are highly reliable, and MEMS can be manufactured from metals using injection molding or embossing. The ceramics can offer a combination of material properties.

For manufacturing MEMS, thin films should be deposited in the range of 1 to 100 micrometers. The deposition process can be physical or chemical. In the physical deposition, the material is moved from the source and deposited on the surface using techniques like sputtering and evaporation. In a chemical deposition, the material is grown on the surface using chemical reactions, and this can be achieved by techniques like thermal oxidation and low power chemical vapor deposition.

Lithography is the process of transferring the pattern by using radiation sources. Various kind of lithography techniques used is Electron beam lithography, Ion beam lithography, X-ray lithography. If the material is resistant to radiation, techniques like Ion track technology can be used.

There are two kinds of etching - dry and wet. In wet etching, the material is removed selectively by dipping into a solution, and in dry etching, the material is dissolved using vapor phase etchant and reactive ions.

Microelectromechanical systems (MEMS), integrates sensors and actuators, on a tiny (millimeter or less) scale. The basis to MEMS technology is a microfabrication technique that is similar to what is used for microelectronic integrated circuits. The primary advantage with MEMS technology is that it allows the production of sensors and actuators on a massive scale at much lower costs. Because of these reasons, MEMS is an attractive option for many IoT applications.

And how mems can enable inkjet printers and airbags?

Thermal bubble ejection is used by inkjet printers to deposit ink on paper, which is essentially the process if printing. The airbags are triggered by the data obtained from the accelerometers in the cars, and they are also responsible for the Electronic Stability Control of the cars.

Summary:

- Micro-Electro-Mechanical Systems or MEMS is a technology used to create tiny integrated devices or systems that can combine mechanical or electrical systems.
- MEMS integrates sensors and actuators in a microscopic scale.
- Components of MEMS come in the range of 1 to 100 micrometers.

8. Standards and Protocols

Preview:
- How is connectivity achieved in IoT?
- Different types of wireless networks

"With the IoT, we're headed to a world where things aren't liable to break catastrophically – or at least we'll have a hell of a heads' up. We're headed to a world where our doors unlock when they sense us nearby."

– Scott Weiss

An essential aspect of IoT is connectivity, as there are multiple devices to be connected. There are several IoT connectivity and network, but in most of the cases, it will be a combination of multiple network protocols. There are different levels of connectivity required in IoT network. For instance, the connectivity between devices in an IoT system will happen at a close range such as in a smart home solution. Whereas, at the same time there will be connectivity required between the edge devices and a central node or a cloud network which requires massive distance communication protocols, for example, a real-time traffic information system. Connectivity standards are also different, depending on the power that is needed and the volumes of IoT data transmitted, adding to the broad range of standards and solutions. Connectivity in the sense of connected devices is the start; connected data is where the value starts.

Based on the area or the range covered, connectivity networks are divided as PAN, LAN, WAN, and MAN. Below described are the different types of communication protocols that are available for the IoT are:

Wireless Personal Area Network:

Bluetooth

Bluetooth is another wireless connection technology for the exchange of data over short distances. Bluetooth is managed by a non-profit organization called Bluetooth Special Interest Group. The latest version of Bluetooth standard, Bluetooth 5.0 has been specifically designed to deliver better support for IoT.

Zigbee

Another short-range wireless communication protocol which is widely used in home automation and the industrial segment is Zigbee. Zigbee is based on the IEEE 802.15.4 protocol and is preferred in small-scale applications with the low data rate, low bandwidth, and low power requirement.

Zigbee is simpler and less expensive than other technologies. Some of the advantages of the Zigbee network include low power consumption, high scalability, security, and durability. Zigbee data transmission ranges from 10 meters to 100 meters at a maximum data rate of 250 kbps.

Significant applications involving Zigbee network include wireless light switches, home energy monitors, traffic management systems, and other consumer and industrial equipment that requires short-range low-rate wireless data transfer.

In IoT, it is essential that the smart devices communicate in the same language at the application layer. Keeping this in mind, Zigbee has created a universal language for IoT called "dotdot," which aims at making possible the communication between IoT devices from multiple vendors across different network layers.

Z Wave

Z Wave is a first short-range communication protocol and operates at a frequency of 900 MHz. Z wave has a little data rate of 100 kbps and is used for home automation which has low power requirement.

, Bluetooth, IEEE 802.15.4, Z-wave, LTE-Advanced, Near Field Communication, ultra-wide bandwidth, Low-Power Wide-Area Network, and emerging standards.

Wireless Local Area Network:

Wi-Fi

Wi-Fi is a wireless local area networking technology which facilitates smartphone, computers, smart TV and other devices to connect with the internet within a particular range, based on the IEEE 802.11 standards.

Wi-Fi technology makes use of radio wave for communication and works best for the line of sight.

Wireless Wide Area Network:

Wireless Wide Area Network consists of technologies such as cellular network technology including 3G, 4G, LTE-Cat M, and NB-IoT and non-cellular low power full area network technologies (LPWAN) such as Sigfox, LoRa, and DASH7.

These are some of the significant communication protocols which have been used in IoT networks. However, most of the networks will have a combination of these technologies are used to cater the respective requirements.

Summary:
- Connectivity is one of the fundamental requirements of the Internet of Things.
- There are various levels of connectivity required in an IoT system depending on the size and scope of the network.
- PAN, LAN, and WAN are three kinds of network technologies used for implementing connectivity in an IoT network.

9. Security

Preview:
- Security concerns in IoT devices
- Types of attacks
- Challenges in IoT

"Security by design is a mandatory prerequisite to securing the IoT macrocosm, and the Dyn attack was just a practice run."

– James Scott, Sr. Fellow, Institute for Critical Infrastructure Technology

In 2016, the Mirai attack exploited IOT devices to infiltrate the internet infrastructure resulting in tremendous shutdowns in North America and Europe causing about $110 million loss. With the increasing user base, this is indeed a great challenge.

The "things" in IoT are vulnerable to external attacks creating complex security issues for IoT. As the number of devices connected increases the complexity increases and it creates more points of entry for attackers. This is because a single compromised device can serve as the launching point to attack other devices and systems. Below are some of the significant security attacks which are a threat to IoT systems. Since wireless networks are used in IoT, the vulnerability of the system is further increased.

Cloud Attacks:

The increasing rate of cybercrime and cloud attacks creates a potential threat to the IoT data stored in the cloud as a lot of IOT data is stored on the cloud

Distributed Denial of Service:

Distributed denial of service makes use of multiple computers to attack a network or a node by flooding it with data and thereby paralyzing the target. The target will either slow down or get crashed, and thereby the remaining valid users will lose the access to the network.

Since there is a large number of devices connected to the IoT network offers a wide target area for hackers. One such incident was the Mirai Botnet attack which almost brought down the internet through a Distributed Denial of Service attack. It is tough to tackle a DDos, because of its distributed nature.

Interoperability

As IoT powers the interconnectivity between devices and thereby create smart homes and smart industries, there are some associated challenges which hinder its expansion. With the increasing number of devices in the network, many a time there will be devices from multiple vendors in the network creating interoperability issues. On top of that, there are too many players in this market today, and each of them is developing solutions independent of one another.

Different devices make use of different platforms and frameworks which makes it impossible to integrate with each other. However, recent IoT standards are trying to minimize this problem.

Privacy

IoT devices collect a lot of data which are specific to individuals. This includes their personal information, day to day activities, location details, health information, and shopping pattern and so on. This raises concern about privacy issues in IoT because in many cases users don't have a control on with whom this data will be shared.

Scalability

Scalability is the capability of a system or a network to adapt to the increasing amount of work or its ability to accommodate future growth and expansions. When the iot network is increasing in size without proper security measure, the scaling would only result in more and more vulnerable points in the system.

Even if the individual devices are secure on its own, when they are a part of a complex network of devices, the most vulnerable point will be deciding the level of security of the entire network. To prevent exploitation of such vulnerable points in the networks, an end to end encryption should be implemented which will decrease the possibility of the network being infiltrated.

The standard which specifies how different parts of the network should interact is still in its nascent stage. This lack of clarity regarding the standards can affect the adoption of iot. Industries can use iot for developing their new innovations, but with the foggy standards, they will be demoralized to do so. The end to end encryptions cannot be developed quickly without having well-set standards.

Summary

- Security is a massive concern in IoT networks as there are multiple access points in the network and breaching into a single device can put the entire network into risk.

- Since a lot of IoT data resides in the cloud, security of the cloud has to be ensured.
- IoT also has other challenges like interoperability issues, the privacy of data and scalability.

10. Future of IOT

Preview:

- A glance into IoT market predictions.
- Opportunities in IoT

"As the Internet of things advances, the very notion of a clear dividing line between reality and virtual reality becomes blurred, sometimes in creative ways."

– Geoff Mulgan

IoT has been a game changer for many industries and companies have been trying to outdo each other by being an early adopter. In this Bull Run, the IoT market is becoming a tremendous growth contributor to the IT industry as a whole. Let's have a look at some of the market figures.

CISCO has predicted that there will be 50 billion devices by the end of 2020. This is a big market opportunity to IT hardware and software vendors and resellers. Most of the IT manufacturers have projected IoT as their principal focus area for 2018.

Gartner has evaluated the IoT market to touch 1.9 ten dollars by 2020. This again opens the market to service providers and vendors associated with IoT.

IoT is not a revolutionary tech on its own, and it has to join forces with other emerging technologies like AI, fog computing and blockchain. 2018 will mark the start of such collaborations. The fog computing can solve the reliability, bandwidth, cost and latency issues. Blockchain will eliminate the need for a centralized authority. The AI can run a real-time analysis of the IOT driven data.

There will be a greater focus on the IOT security since IOT related cybercrimes are expected to spike with the adoption of the technology. The concerns industries will be taking care of security at every stage of Implementing IOT right from the deployment.

There will be a corporation model coming up with the adoption of IOT, where many companies will create an ecosystem to innovate and develop IOT solution together. This will result in large-scale IOT projects based on open source codes and open architecture which will be cost effective. With this cooperation culture in place, there will be shift from the single-vendor approach. Even the customers can get involved in this community so that their needs are better catered.

Till now most of the focus of IOT was on automating everything, but with its upcoming significant convergence with other technologies, it can potentially invent a new business model and new kinds of value additions. With new business models in place, the industries will be flourishing like never before. Even with the open and cooperative structure integration will be a big challenge.

Summary:

- CISCO has predicted that there will be 50 billion devices by the end of 2020.
- IoT is opening up a vast potential to hardware and software OEMs in the IT industry.

CONVERGENCE

1. Convergence of AI, IoT & Blockchain

"Isolation is a blind alley...Nothing on the planet grows except by convergence."
-Pierre Teilhard de Chardin

"With better technology comes convergence and in many cases replacement."
  - Tom Edwards

Convergence of Ai, IoT, and Blockchain

As we are moving towards a rational world, where things can provide services autonomously, it is important to converge the technologies in a way that it resonates and work together as a system for the betterment of the future. The technology stack for the future will start with IoT generating the data which will be authenticated, validated and secured by Blockchain systems and then processed, analyzed and automated using Artificial Intelligence.

Connectivity is the fundamental force behind shaping technology today as the way it is, and it is the lifeline for emerging technologies such as AI, IOT, and Blockchain. Even though, the scope of connectivity is increasing still there is a disconnect in the way we are connected.

The World Wide Web has a significant role in achieving mass adoption of the internet. It has evolved throughout the past two decades and has helped to bring the world closer together irrespective of the physical distance between places, but at the same time, it is getting centralized to a great extent which is not advisable for many reasons. Complex requirements of today and the future cannot be met with a centralized architecture which creates silos and will not work together effectively. To understand where the future is going it is necessary to understand the evolution of the web.

2. Evolution of the Web

Preview:

- The journey of the Internet
- How did the web evolve?
- Web 1.0 Vs 2.0 Vs 3.0 Vs 4.0

"History is not just the evolution of technology, it is the evolution of thought" - James Redfield

Web 1.0

Web 1.0 refers to the initial stage of the World Wide Web, and they were made up of static web pages without any interactive content. It comprised of read-only websites, which had catalogs and brochures similar to magazine and newspapers. It was also referred to as monodirectional websites, and the primary objective of these websites was to broadcast information and create an online presence for businesses.

Thus Web 1.0 just allowed people to be the consumers of the content. The core protocols used in Web 1.0 were: HTTP, HTML, and URI.

Web 2.0

Tim O'Reilly coined the term Web 2.0 in the year 2004. He defines Web 2.0 as:

"Web 2.0 is the business revolution in the computer industry caused by the move to the internet as platform, and an attempt to understand the rules for success on that new platform. Chief among those rules is this: Build applications that harness network effects to get better the more people to use them."

Web 2.0 consisted of people-oriented websites and emphasized on user-generated content which was easy to use and can work well between multiple systems or devices. Web 2.0 websites became interactive, and it made the web bi-directional. In contrast to Web 1.0 which had websites which were passive, Web 2.0 incorporated platforms which allowed people to collaborate and interact with each other by creating the content rather than just consuming it.

Web 2.0 had a flexible design and collaborative content creation compared with web 1.0. The significant additions that came along with Web 2.0 were social networking sites and social media sites like Facebook, blogs, Really Simple Syndication tool, wikis, video sharing sites like YouTube, hosted services, Web applications, collaborative consumption platforms, and mashup applications.

- Blogs- Blogs are dynamic web pages which are informative and interactive and can be created by individuals or a group of people or for businesses. Blogs constitute regular posts which have to be updated by users in the form of texts, photos, videos, etc.

- Simple Syndication - RSS is a family of web feed formats used for syndicating content from blogs or web pages. RSS is an XML file that summarizes information items and links to the information sources. Using RSS, users are informed of updates of the blogs or websites which they're interested in.

- Wikis- A wiki is a web page or a set of web pages that can be easily edited by anyone who registers onto it. Unlike blogs, previous versions of wikis can be examined by a history function and can be restored by a rollback function. Wiki features are included: wiki markup language, simple site structure and navigation, simple template, supporting of multiple users, built-in search feature and simple workflow.

- Mashups- Web mashup is a web page or a website that combines information and services from multiple sources on the web. Mashups can be grouped into seven categories: mapping, search, mobile, messaging, sports, shopping, and movies. More than 40 percent of mashups are mapping mashups. It is easier and quicker to create mashups than to code applications from scratch in traditional ways; this capability is one of most valuable features of web 2.0. Mashups are generally created using application programming interfaces

Web 3.0

Web 3.0 is also referred to as the Semantic Web, and it is data-driven. Data is the key in Web 3.0, and this data will be collected from the users of the web and thereby adjusting the web according to the needs of the users. An example of this is targeted advertising, where you will get an ad if you search for something.

Two things form the base for Web 3.0: Semantic Markup and Web Services. Semantic markups focus on reinforcing the meaning of a webpage or web application rather than focusing only on the presentation. Web services are software designed to support node to node interaction in the network.

Web 3.0 also focuses on increasing the accessibility of the internet to the people who are mainly driven by the advent of mobile and cloud apps.

Web 4.0

Web 4.0, also referred to as symbiotic web is in its nascent stage today and will grow along with the emerging technologies like artificial intelligence, IoT and

Blockchain. This exactly is the opportunity lies. Symbiotic web combines data science and analytics with web technologies and thereby make the web intelligent. It enables machines to analyze the feelings and emotions of the users and to optimize the interaction between humans and computers. Web 4.0 will enable building mind controlled interfaces which would analyze the contents of the web as well as user-generated data, and react efficiently. Facial recognition technologies can also be used in figuring out the emotions of the user.

Thus Web 4.0 is an open, linked and intelligent web which requires the convergence of emerging technologies such as IoT, artificial intelligence and blockchain. The future awaits an intelligent web which will be comparable to the human brain.

Summary:

- The World Wide Web evolved over a period to connect the world as the way it is today.
- Web 1.0 consisted of the static website and was mono-directional.
- Web 2.0 made websites more interactive with user-generated contents and incorporated features like blogs, wikis, RSS, etc.
- Web 3.0 focused on the semantics of the web and emphasized in adjusting the web according to the user needs.
- Web 4.0 aims at creating a symbiotic web which is intelligent and making the interaction between humans and machines more real. Web 4.0 will converge technologies such as AI, Blockchain, and IoT in its implementation.

3. World of Centralized Data

Preview:
- What is data centralization?
- Drawbacks of a centralized system?
- Measures for decentralization

"The old world is dying, and the new world struggles to be born; now is the time of monsters."

Antonio Gramsci

As we have been discussing throughout the book, a huge amount of data is being generated day by day, and with increasing number of IoT devices, there is an exponential growth in the amount of data. Further, a lot of personal data is being generated by social media websites, e-commerce websites and so on. In a centralized network, data will always remain in silos and will give additional authority to a few organizations. Today, companies like Facebook, Apple, Amazon, Google, and Microsoft generates a bulk of user data which they store centrally in their network, and as a result, artificial intelligence system gives these organizations power to process this bulk data. These organizations create a barrier to the entry of new players in the market by controlling the huge amount of data.

Another growing concern is about the privacy of this data. We have seen how Facebook was linked with a data privacy breach with Cambridge Analytica, a political consulting firm who used data to change the behavior of the voters. Further, a centralized network will not be scalable for the next generation were billions of devices are going to generate an enormous amount of data. The ultimate result of the centralization of data will be the existence of organizations and authorities who will levy substantial capital benefits from the common man.

What if we can trust a software code that can put an end to these centralized systems? Yes, that is the convergence of IoT, and AI with blockchain begins. It is the roadway to the deep web creating a decentralized system and will provide provisions for the rightful owner to control and share data. It will also create new business opportunities, especially in the service sector. Thus, the fact that AI, IOT, and blockchain are complementary creates a massive opportunity for businesses which have to be appropriately leveraged.

So are we going to have a decentralized, secure and intelligent system tomorrow? The answer is no. We have already seen how the internet was received in the initial days. People were confused whether it is going to benefit the society. However, today breaking all the barriers we have Internet bringing the world closer together, and today if the world is not connected to the web it might bring the globe to a standstill. Similarly, the convergence of these technologies will result in a better system in the future. Let's hope that will be short.

Summary:

- With the number of IoT devices increasing day by day the amount of data generated is exploding.
- There is a growing concern in the centralized storage of this data being generated, because of multiple concerns like performance issues, security, and privacy concerns, etc.
- The centralized system gives additional power to a limited section which has the authority over the data.

4. Need for decentralization

Preview:
- Why do we need decentralized networks?
- How will blockchain secure IoT networks?

"I'm not going to stop the wheel; I'm going to break the wheel " Game of Thrones

Need for Decentralization

Connectivity is the basis of technology today and data has to be shared in real time. The evolution of the web and emerging technologies complement each other and will have a profound impact on the way things work.

However, there are some challenges with the web. With a large amount of user data like personal data and behavioral data, it can raise considerable concern about privacy and security. Further, the web architecture is not yet ready for the next generation of internet computing of data explosion. As discussed above, Centralization of data is another major concern and data is getting centralized towards some of the companies which helps them to polarize decisions and take actions which are beneficial to their stakeholders.

With IoT coming into picture devices are extensively getting connected to the web and enabling new business models. These devices generate a large amount of data which have to be enhanced using AI and machine learning algorithms. This might consist of data which requires privacy and security and some data might only have to be shared with some trusted parties. Thus, IoT devices are exposed to threat from the endpoints to different layers of the network till the cloud or the central nodes.

Further, as machines are starting to interact and transact with each other without human intervention or any other intermediaries, we have to develop a basic level of trust that the machine is robust, resilient and it is protected from external threats or malicious attacks. It has to be ensured that it is self-sufficient and capable of doing the right things securely.

How do we do this? Whom should we trust and how to establish the trust?

Enabling trust with Decentralisation:

Thus, there are many challenges with centralized networks, like scalability and security issues. With the growing size of the network and the pace of data and connectivity requirements with the emerging technology, it is necessary to explore the architecture which can suffice the demands of the future requirements.

Peer to Peer technologies has gained much popularity because of its ability to improve performance, scalability, and efficiency of the network. Unlike the client-server architecture peer to peer networks doesn't have a single point of failure and makes it difficult for external attackers to filter through. That is where blockchain becomes beneficial to IoT and AI.

As discussed in the chapter, A guide to Blockchain, The most significant advantage of blockchain technology is that it can induce trust in the form of Software code thus eliminating the requirement of human intervention. So what if blockchain can enable trust when machines are transacting with each other? Blockchain works with decentralized consensus algorithms which approves a decision (it can be a transaction of money, property or information which involves the addition of a new block to the historical ledger) only when the entire nodes in the network validate it.

As you already know, the journey of Blockchain started with bitcoin which is a blockchain protocol for transacting monetary value. With Ethereum introducing smart contracts opened the way to a transparent and conflict-free transaction of anything of value utilizing a distributed and decentralized network of Ethereum Virtual Machines.

Summary:

- Decentralised technologies can impart security and trust in IoT networks.
- Entirely autonomous systems will require integration of technologies such as AI, Blockchain and IoT.

5. Integration of AI

Preview:

How IoT needs AI and Blockchain?

What is the level of security and authenticity of the data generated from IoT devices?

"We will continue to see a convergence of the digital and physical world, Those who conquer it will be market leaders."

-John Phillips

We have heard of all sort of smart things; they are supposed to show some intelligence to meet our expectations, they should be able to respond to specific situations. The IOT enabled objects are supposed to catch data associated with the objects to react to the situation or to communicate with other devices. These activities are bounded by the programming they are infused with. This is where AI comes in and increase the scope of the situation by giving the objects the ability to learn and respond even when the environment is changing. Autonomous cars use IOT sensors to understand the environment using computer vision and predict the routes using interwoven GPS networks just like the google maps which can dynamically set routes and adapt to changing traffic conditions.

When any malfunctions or security breaches occur within the system, the IOT devices will be responsible for the trigger which can solve the issue. Prevention is always better than cure, using predictive analysis the issues could potentially be solved before it even occurs. The system will be much more than a response to situation environment by being an adapting and learning unit. The triggers can call for action which can solve the situations automatically or semi-automatically based on the gravity of the situation.

There should be an incentive for the people to contribute data. Of course, there will be better products and services catered to them based on those data, but short-term benefits will be more compelling for data contribution like the data contribution being monetized using some applications, this raises a lot of questions like how can the ownership of the data be validated? Who requires the data? How can the value of data be decided? How can we safely provide access to data? How can we make sure there is enough consent for the usage of this data? Our traditional applications can't answer these questions; this is when blockchain comes in which will ensure the transparency of the system by maintaining the record of data which can be traced back to its roots and audited very easily. You cannot falsely claim the ownership of data in a blockchain powered network as the lifetime of each data is timestamped in the network such that the real owner will benefit from any data transactions associated with him. Common protocols can be preset using smart contracts; the network will itself ensure that the conditions are satisfied and thus providing the network with the top-notch database.

The decentralized nature benefits IOT security greatly, the IOT consumers are not ready to spend more on security in the current situation, and the providers are hesitant to provide security services free of cost. This is mainly because of the business model of IOT industries which expects the customers to deal on an upfront payment basis. Adoption of new business models can solve this situation, by creating a win-win model for the consumer and the supplier. The supplier should be incentivized for additional services. Predictive analysis can be used to motivate the customers to invest in security, to prevent a more significant loss. The transparency and incentive system can be implemented only by using distributed ledgers and smart contracts.

Summary:

- IoT generates a lot of data which will be underutilized if we do not implement AI systems to learn the data and take actions accordingly.
- Further Predictive Analysis can be used within IoT networks so that threats can be predicted and encountered even before it happens.
- Further, the blockchain network enhances the security and efficient utilization of data.

6. AI needs IOT

Preview:

- How AI needs IoT and blockchain?
- How will data be used in the future?

"When we talk about the Internet of Things, it's not just putting RFID tags on some dumb thing, so we smart people know where that dumb thing is. It's about embedding intelligence, so things become smarter and do more than they were proposed to do."

– Nicholas Negroponte

We already know those fancy entrances with access enabled by IOT devices. These are mostly used in office spaces so that the entry to different regions within the office can be maintained in a hierarchical order. If the information from these devices is analyzed using artificial intelligence, the time spent by a person in a particular region and the movement patterns can be recognized. With such data, the office can be designed to maximize efficiency. Even any kind of suspicious activity can be tracked with the help of AI; this can be achieved by using predictive analysis as well which is getting improved with time. The issue with the predictive analysis is the possibility of bias which could frame wrong people or put innocent people under the radar.

If facial recognition and emotional analysis are run on a large scale, the marketing and selling of products will be much easier. Using IOT devices the time the users spend in front of a particular product can be tracked, with proper facial recognition system we can analyze the emotional status of the person, and thus understanding to what degree they prefer the product. Already when we are visiting websites, our mouse pattern movement is analysed to understand which products we prefer and what portion of the site interest us the most, using this data we are given custom-made content in order to maximize the presence of things we prefer which will benefit them as well as it can increase their sales by presenting us more preferable objects. The preference patterns can be calibrated and modified using AI.

When equipment breaks down, it can affect the profit levels, especially in manufacturing industries. We can maximize the efficiency of the industries by minimizing the downtime involved with faulty equipment. This can be achieved using IOT sensors derived data and using it for detecting faults by running predictive analysis using artificial intelligence. The equipment will be attached with sensors so that the vitals of the machinery is updated which can use to approximately find the breakdown time so that the machinery can be repaired or replaced as soon as the time approaches, which will prevent the chaos and confusion around the event when it happens unexpectedly.

In short, wherever sensors can be used to record useful variable information, the combination of that information can be used by artificial intelligence to derive useful correlation and projections which can benefit the working of the concerning industry greatly. When a particular industry is concerned, using machine learning can improve the process more as the time progressing as it gets more and more accustomed to the industry.

So, it is clear that to get full value from iot devices; artificial intelligence is necessary otherwise iot devices will be just a source of vast chunks of data which will keep on accumulating with no further applications. Humans are known to understand pattern which machines can't, but when we are dealing with this much volume of data, it is practically impossible for humans to get a proper exposure so why IOT and AI needs each other.

Artificial Intelligence has succeeded dramatically in exhibiting a narrow spectrum of intelligence like in case of face recognition, chatbots, spam filters and all sort of necessary natural language processing. We have experienced the chatbots which can imitate human assistants to a great extent. These AI application can sometimes exhibit beyond human abilities

AI can be smarter than the human in recognizing faces, don't we fail to recognize some people at times even after seeing them for a couple of times. Talking about Virtual assistants, we have Siri, Alexa and all which are pretty good at many things like playing our favorite songs on demand.

So what's, is stopping the AI to overall function better than the human brain? The lack of computational power. Also the better we understand the human brain easier it will become to replicate its activities. According to Anders T Sanberg and Nick Bostrom of the Future of Humanity Institute, emulation of the human brain will help us set clear-cut goals for computational neurosciences. The validity of several assumptions can also be made clear with the newly set goals.

The neural networks which were created were severely lacking computational power, memory and storage specifications. It will take months with the currently available processors. The entry of quantum computing can solve this. There is a lot of investments flowing for improving the chips so that it can replicate the activities of the human brain. The power efficiency can be increased when the brain structure is used because specific activities can be dedicated to different sections of the system. This is where IOT and Blockchain come in, to help AI distribute the functions to different sections to improve the efficiency. The decentralized nature of the blockchain technology can help AI to make several units function towards a shared goal. Currently, AI can't ensure that consensus can be reached in a distributed network, which is a particular case of Byzantine problem which we have discussed in detail in the dedicated chapter. So using blockchain, we remove the trust and make the system of AI units trustworthy. Even if a third person controls the swarm of AI units, because of the transparent nature, any anomaly can be quickly dealt with.

We know the power of data, better the data we have, we have can better results by feeding into our AI resources. There is a high demand for refined and quality data so that products and services can be tailor-made and sold to the public. With real-time IOT services, it can be ensured that the standards are met.

The entire system can be designed such that every participant can be rewarded using crypto economics powered by blockchain. The flow of data can be rewarded so that there will be a competition to provide better data which will self-improve the quality of the network

Summary

- AI and predictive analysis are going to change the future of many industries.
- Data generated from IoT devices play a key role in designing these systems.
- Further blockchain networks can help AI systems to work as decentralized units and to achieve consensus among different units.

7. Fetch, OP and IOTA

Preview:
- Companies converging AI, IoT and Blockchain
- What are their objectives?

How do they achieve this?

Technology will not replace great teachers, but technology in the hand of great teachers can be transformational - George Couros.

i. Fetch.AI

Fetch.AI is a next-generation protocol which extrapolates the capabilities of IoT, AI, and decentralized technologies like Blockchain and Decentralized Acyclic Graphs. It was founded in 2015 by a team of AI and Machine learning pioneers of Deep Mind which is a part of Google Alphabet. Fetch.AI mainly delivers 2 things: a digital currency and an intelligent, decentralized transaction protocol.

The objective of Fetch.AI is to give a new face to the world economy by creating a collective superintelligence which enables artificial systems to transact and interact with each other without any human intervention. Fetch actively identifies a requirement with its intelligence and makes it meet with another value generating agent that can offer this requirement. This means that a machine will be able to buy or sell digital assets to another machine with contracts, payments, and execution, everything handled autonomously, without anybody else's input.

For example, you can build an intelligent agent to trade excess of storage or compute to other agents in a fetch network where there is a requirement for the same in real time and all the associated processes like payments and execution will be autonomously carried out.

Let's understand in detail how this collective superintelligence is achieved. The collective super intelligence of Fetch is achieved through three technology layers:

Layer 1: Autonomous Economic Agents

Layer 2: Open Economic Framework

Layer 3: Fetch Smart Ledger

An autonomous agent is an intelligent agent operating on behalf of its stakeholders without any external interference. The Autonomous Economic Agents (ASA) in Fetch are software codes which can perform actions like delivering data or services on its own without any external input. An AEA should have a unique identifier which enables them to transact using Fetch Token. Further, every AEA will have to maintain the details of all the other nodes they are registered with. Given below are some of the different types of Agents.

Inhabitants: They are AEAs integrated into a physical object or device. For example, AEAs in drones, mobile phones or cameras.

Interface: AEAs that acts as an interface between the current economy and the new economy.

Pure Software: AEAs which exist only in the digital space.

Digital Data Sales Agent: A specific type of simple software AEA focused on extracting value from data. This intends to solve one of the major problem in the data industry: "data doesn't sell itself.."

Representative: AEAs are representing the owner or stakeholder of an entity. This AEAs have algorithms to learn the preference and tolerances of the stakeholder and thereby acts accordingly.

The AEAs lives within the second layer, Open Economic Framework. It dynamically reorganizes to create the best digital world for the AEAs, that is, it positions the agents in digital space such that it can offer values to the agents that are searching for that value.

The Fetch Smart ledger is a next-generation digital ledger and is a unique combination of Blockchain and DAG architecture so that to bring the best of both the worlds. The issues of scalability and limited transaction per second of blockchain are resolved in Fetch since it has to support a large volume of low-value transaction.

The Fetch Smart Ledger creates a collective market intelligence by self-learning and thereby restructuring itself to optimize the Open Economic Framework.

ii. Ocean Protocol

Ocean Protocol is a decentralized network for data exchange and acts as connections between the data providers and consumers of the data. The objective of Ocean protocol is to make more data available for AI systems and thereby to optimize and to harness its full potential.

With an increasing number of smart devices, the amount of data generated is increasing exponentially. The total amount of data generated across the world in 2010 was 1 ZB, which increased to 16 ZB by 2016 and is expected to reach 160 ZB by 2025. But, according to Mckinsey, only 1% of this data is effectively utilized for analysis. This is one of the challenges which Ocean protocol is trying to solve.

Data is exceptionally crucial for AI systems because it requires data to become accurate and it optimizes as the number of data increases. Ocean helps to unlock this data, especially for AI. Ocean uses blockchain technology that allows data to be shared and sold in a safe, secure and transparent manner.

Ocean protocol has a tokenized ecosystem and rewards service providers in return for AI data and services. The ocean network comprises of different stakeholders who offer specific values to the network and gets paid for it with Ocean tokens. Given below are the key stakeholders in the ocean ecosystem.

Data or Service Provider: These are the people who have AI data or services in their custody and can supply it over a marketplace for which they will be paid in Ocean Tokens.

Data or Service Consumers: These are participants in the Ocean Network who can consume data or other services for their use for which they have to pay using Ocean Tokens.

Data or Service Curators: This includes exchanges, marketplaces and other application providers. It facilitates the interaction between providers and consumers.

Keepers: They support ocean protocol by running the nodes within the network. They are the miners of the blockchain network.

Since ocean protocol is a permissionless protocol, anyone can run an ocean keeper node and anonymously participate in the network. However since this technology is at its nascent stage, some challenges and concerns are revolving around.

iii. IOTA:

IOTA is an open source decentralized distributed ledger technology with the objective to power the Internet of Things with feeless microtransactions.

IOTA tries to overcome the scalability and a limited number of transactions per second issues of the blockchain, and therefore uses another distributed ledger technology called Direct Acyclic Graph (DAG). This distributed ledger of IOTA is called Tangle, and it does not require miners to validate transactions, and therefore the transactions are ultimately free.

Unlike a blockchain ledger, Tangle does not group transactions grouped into blocks and store it in sequential chains. Instead, they are stored as a stream of individual transactions entangled together. Further, there is no hierarchy of roles in the network and everybody have an equal role in the network. It also makes use of a different mechanism for validating the transactions without miners. To make a transaction in the network, one has to help in the validation of any two previous transactions in the network. After this, your transaction will be validated by some subsequent transaction.

Thus IOTA does not rely on financial rewards, and as more people start using the network, the network will become more fast and optimum. The two main features which make it a choice for the machine to machine transactions are:

Micro-transactions:

A larger number of transactions per second:

The major objective of IOT is to become the underlying protocol for the internet of Things.

Summary:
- Fetch.AI aims at creating digital entities that can transact independently without any human intervention and can represent themselves devices, services or individuals.
- Ocean Protocol is a decentralized network for sharing data and associated services. The objective is to make more data available to AI systems.
- IOTA is another distributed ledger technology that aims at becoming the underlying protocol for the Internet of things. It aims at powering feeless microtransactions for IoT network and uses a different technology called Distributed Acyclic Graphs.

APPLICATIONS

1. Advertising

Preview:

- Enabling Intelligent and dynamic advertisements
- Outdoor advertisements that can track exposure

"Never stop testing, and your advertising will never stop improving."
- David Ogilvy

With the new business models emerging with new technologies advertising is sure to get intense and exciting. When outdoor advertising is considered, it is getting more and more digitized. What can these digital advertising boards do?

With the use of cameras, image processing, and movement tracking sensors, we can calculate the amount of exposure each advertisement is getting and the parts of advertisements which is getting focused can be analyzed to create more efficient and dynamic advertisements. It is said, conversions from outdoor advertisements can't be calculated because they are mainly used for brand awareness, but sure we can increase the awareness by maximizing the exposure factor.

Depending on the size of the advertisements, facial detection, and emotional responses can also be recorded and analyzed to understand the customer requirements and the rough estimates of customer conversion rates.

Each advertisement spot can be tagged with the previous client data, the transparency of which can justify the price of that spot, this obviously can be accomplished using the blockchain. The immutability does pay off here. The numbers cannot be manipulated and forged so that the ad spot is sold. Even the exposure factor can be tokenized such that ad will be displayed in the platform till the tokens are exhausted.

With this system in place, you will have to select the target audience using the respective application which will work somewhat like the facebook ads. You have to specify the target audience and the required exposure, rest will be accomplished by the AI which will find you the best possible packages with minimum usage of money. For example, if your target audience is students in California. The campaign will initiate your ads in the ad spots nearby universities in California like CalTech, and the selection of spots will be made so as to match your budget effectively. The advertisements will be cycled such that maximum coverage is obtained, that is you just paid for your exposure tokens, these ads will be dynamically adjusted to the tokens are exhausted.

This process is much depending on how the exposure value is calculated; the implementation will require a large-scale adoption since the proper optimization and exposure can only be obtained on a large scale. The application on the client's end will be receiving the statistics of the ad regularly using the sensors attached to the device so that the client will feel that his investment is adequately justified.

Another possibility with the future of outdoor advertisements is the application of direct call to actions of course depending on the size of the concerned advertisements. This feature can be implemented using 2 step security processes so that spamming can be avoided.

The advertisements can be made to react to your proximity assuming the location permissions are provided, for which the people could be incentivized in some way so that the location info for more effective advertisement. One option is to tokenize the location data, depending on the interaction of the devices with the ad boards, the people can be rewarded with tokens of the advertising network, which could potentially increase the participants in the network. As location services are battery intensive at least, for this to become the more efficient process, our mobiles devices have to become more efficient for the frequent use of location services.

With all this in effect will we be able to experience better products tailored for us or will it look more like privacy invasion?

Summary
- Cameras, image processing, and motion tracking sensors can be used to create an exposure index for offline advertisements
- The history of outdoor spots can be tracked with the immutable distributed ledger.
- Call to action can be implemented in the dynamic advertisements.

2. Agriculture

Preview:
- Automating the process of agriculture
- Benefits for the farmers

"To make agriculture sustainable, the grower has got to be able to make a profit."
- Sam Farr

Conservation of water, reducing levels and controlling fertility levels are critical factors concerned with agriculture. There are sensors available which can be used to detect the level of moisture and nitrogen. When an everyday automated watering and the fertilizing system is used, they are applied at particular intervals which is predetermined and also not the best practice. With the data from these sensors, direct correlations to the yield amount and quality can be derived

Weather predictions can also help the farmers to adapt faster to the changes which are about to come. There are applications dedicated for weather reports exclusively for farmers. Imagine the losses you could avoid if you are prepared for the upcoming hail storm or other weather extremities. The field can be overlooked by satellite imagery without the presence of the farmer himself. These images are only going to get more and more accurate with time.

We often don't know where our food items come from, even if we are considering a packet of uncooked rice we don't know about the origin and the processes it went through in proper clarity. If you are wondering what the point of tracking the timeline of a rice pack is, the price of the packet can be justified by keeping a record of the processes involved. If you are buying a high-end product you should feel that you are not paying those extra bucks for anything, you should feel the value in the record attached to the packet.

This tag carrying the record of the product's history is implemented using the blockchain. When a crop batch is planted, the tag is created, and the details like date of planting are recorded. The details of fertilizers, pesticides and watering habits are all mentioned and timestamped in the tag. The beauty of it is everything happens automatically, no direct human intervention is required. The machines plant the crops by using available data to space and orient them in the best possible manner. The sensors pick up the moisture and nitrogen levels. When the levels are low, automated machines will water and fertilize the plot. The system keeps on improving and adapting by using the data available from these sensors and the weather information. After the crop is harvested, the details of packing and shipping are included in the tag till it ends up in the shelf of the store. Now you know everything about the product just by scanning the tag. If you are buying something organic, you can be sure that it is indeed organic and not some fake marketing strategy.

The sensors attached to the farming machinery will communicate the data to the AI, which could determine any inefficiency or chances of failure of the machinery. If any anomaly is found, the system can automatically call for a replacement or repair as required. Since this process requires external interference, the farmer can devise the system so that he could confirm the process. The use of sensors and running analysis of those data for better farming practices is known as precision agriculture

AI can also be used for adaptive programming drones which can plant, nurture and harvest crops in areas which are hard to access. Even the quality of these drones can be ensured by using blockchain tags on these drones so that the farmers can make sure they are getting adequate machinery. The drones will also be programmed to use pest control as and when required. AI can also optimize the crops to be harvested, land allocations to different crops and the time gap between harvest and replanting for maximum efficiency

The blockchain implementation also removes the middlemen from the traditional farming system which makes sure that the middlemen won't be siphoning away the majority income of the farmers. That doesn't mean the middlemen will be left stranded. With the emergence of new technologies, there will be endless opportunities which can promise you a better lifestyle. The farmers with the increase in profits from the direct sales can invest more in their machinery and associated software which will improve the quality of the processes even more.

These farming field can even function like closed ecosystem just like the self-sustaining terrariums we can create. But this thought can trigger some existential crisis thoughts in us!

Summary:

- Farmers reap the maximum benefit from the process as the intermediaries are excluded
- The sensors in the plot detect the lacking resources so as it is refilled automatically.
- The entire lifecycle of the food from purchase if the seeds for racking in a supermarket are recorded.
- Intelligent agriculture machines which can self-order for repair and replacement.

3. Autonomous vehicles

Preview:
- Level of adoption required for autonomous vehicles
- Advantages of autonomous vehicles
- The challenges faced

"Everything that's transported over the ground, whether it's people or parcels, will be impacted by autonomous vehicle technology. "
-Karl Iagnemma

Driving a vehicle can be seen as an entirely straightforward process " taking people from a point to another using the best path possible." The problem statement itself sounds like something which can be solved by artificial intelligence and be completely automated.

A mix of the robotic and manual driver can be quite chaotic. If a particular section of society is wholly transported automated, the process will be much more comfortable than when the situation is hybrid. When only free transportation is present, the vehicles can communicate among themselves so that the AI decides the best possible outcomes. The chances of traffic jams can be minimized when the machines can optimize the routes and timing. The robotic drivers can be programmed to respond to the unexpected situation at a faster rate than humans. We already have maps which optimize the best routes and directions for us, which can be used by the robotic drivers to reach the destination efficiently.

The autonomous cars will reduce the cost of transportation, increase safety and mobility. There will be a marginal decrease in accidents as most of the things happening in this environment can be predicted by AI. We won't be staring at the red traffic lights like we are doing now because the AI will be optimized to reduce the waiting time in case it is required. The autonomous cars are highly beneficial for the people who can't or are not permitted to drive. Another issue we face regarding transportation is the parking space, and the autonomous cars will also automatically choose appropriate parking spaces without any human intervention which indeed is a great relief because extra effort finding a parking spot is eliminated. When the journey is optimized, it also means less fuel consumption. Therefore a marginal saving on fuels will be happening if we are considering the entire automated community.

In spite of all these advantages, there are concerns like unexpected passenger interventions on the journey, the degree to which safety can be optimized by these cars, privacy and security concerns like what if the control of the cars are taken over by hackers.

The major challenge of materializing autonomous cars is to manufacture a system which is capable of collecting data through sensors and analyzing to provide all the mentioned benefits. All the objects in the path have to account for which is a complex challenge. A deep neural network is used for solving this issues so that the cars decide the best options given the current situations.

Summary:
- For smooth functioning of autonomous vehicles, the specific area should entirely adopt autonomous vehicle.
- Autonomous vehicles can reduce transportation cost, increase safety and mobility
- Highly beneficial for people who can't drive or are not permitted to drive.
- The accuracy and coordination of sensors is a great challenge.

4. Content creation

Preview:
- Identifying the reading and viewing requirement of the audience
- Enforcing copyright using the blockchain

"If it doesn't sell, it isn't creative."
-David Ogilvy

How can you create dynamic contents so that the viewers are engaged?

The answer is machine learning. A particular content can be split into various segments depending on the information and emotion it is conveying. By using facial patterns and emotion recognition, the AI can analyze how much each segment of the content was keeping the user engaged. The aim of people who rely on content creation for income wants to find an optimal marketing strategy for their content. The AI after machine learning the pattern of content and the reaction can provide us with a marketing strategy which will efficiently cover under the budget. The AI can also mention the structural changes in the content, which can keep the audience more engaged, it can also suggest addition and omission of elements this suggestive technique will be applied after the machine analyses the interaction of a lot of such articles with the people.

If you are wondering if the art of creating contents will die out, when AI is doing all the work, which is not the case. We decide the essence of the content. We decide the hope, expectation, and appreciation that the viewers should experience. The machine optimizes the content such that those ideas are conveyed in the best manner possible.

Websites can presently analyze the pattern of mouse movements and scroll patterns to deduce our level of interest in the different segment of the website. This process is greatly helping the content creators and designers to optimize the website better. Imagine the accuracy which this process can reach if facial recognition is also included in the analysis. The content can also be dynamic such that the creator will present a basic skeleton which will be modified according to the taste and style of the reader.

Blockchain-based platforms like Steem is a content creation platform which incentivizes all the participants in the network - the developers, creators, and curators. Even the participants who comment and like the posts are rewarded. Such a system can provide a positive environment because of the reward system surrounding the network. The toxic mentality of people will be reduced because in the end the majority of the people like rewards and people don't want to be reported and their reputation value to be reduced.

Blockchain can also enforce copyright in a far better way. Once the copyright is recorded in the blockchain network, it will be immutable and secure. Whenever any part of the content is reused, the copyright claim will be automatically invoked which force the user to either will take down the content or to pay a royalty. The payment system can set on smart contracts such that the content is taken down automatically if the royalty is not paid in time.

So in future using this tech you create optimized contents fit for the best marketing strategies which will keep the audience most engaged with security ensured on top of all.

Summary:
- Image processing and computer vision can be used to analyze contents, which can be machine learned to create what the audience needs.
- The immutable ledger with smart contracts will ensure the enforcement of copyright privileges like claiming royalties or removing contents.

5. Education

Preview:
- Creating tailor-made education system for each

"Education is the passport to the future, for tomorrow belongs to those who prepare for it today."
-Malcolm X

Most of the current education system is not tailor-made to fit the requirement of students. A set of mandatory subjects are mostly inflicted upon the students who are rarely motivated to study those subjects. The education system should change such that it is not something that conditions us to function in society without the capability of questioning. The fun part is that the system is succeeding greatly in that.

There will be endless resources which could be accessed by the people at their comfort

The tailor-made curriculum is another excellent attraction for future education. The pattern of the students will be analyzed to provide them with dynamically updated materials, Like the present situation we won't be obliged to follow the same curriculum set at the starting of the course, the dynamic course structure will make sure, maximum efficiency is achieved when the structure is being modified throughout the duration of the course. One of the methods of achieving is by taking the feedback from the student after each section is explored, facial analysis, time spent on a particular portion and emotional recognition. The regularly updated combination of these factors would help to create a tailor-made profile for each student so that proper educational structures and tools could be customized

Evaluation will not be based on multiple core exams and projects. The present situation of examination is just forcing the majority of us to push in knowledge during the last few days before examination only to be forgotten soon. As mentioned earlier, the process of profile creation based on the studying pattern will itself be the evaluation. The study and the examination will be a single process. From machine learning, the data retention can also be analyzed which will determine the right skillset and resourcefulness at a particular time. Also, the time required to refresh the subjects could be analyzed. All this information will be displayed on the interface provided to the student, the information in this interface is recorded using blockchain so that any kind of manipulation is impossible and the data is directly fed by the analysis done by AI through various inputs received from the sensors. The profile can be encrypted using cryptography so that accidental leak of information is avoided. The profile can be shared on request such as an application for jobs or validation for a particular certification.

If you are wondering if this system is confining us to the walls of our houses and preventing any social interactions that are meant to happen at schools, well that is not true. The profile will also track the practical aspects of the curriculum. There will be centers specially dedicated to the practical approaches, which will be accessible according to the current profile status of the student. These centers will serve as the source of social interaction. The practical labs will be backed by IOT devices which would update the data of the student when the task at hand is accomplished.

The students will be very much involved in forming the curriculum which earlier was under the sole decision of the authorities which did not care much about the individual existence of people. When education systems were created, the primary purpose of it was to create a group of people who can follow norms and work without questioning the authority of the system. This has worked so well, and this emerging technology will break the pattern. Our true passions and desires will not be left unseen or ignored with such a system in place. We will no longer be forced into something we are not interested in. We will not be dwelling on subjects till we convince ourselves that we are interested. We will be appreciated for what we like. This is the revolution of education.

Summary
- Education can be measured by the journey of learning rather than being measured by rather shallow tests.

6. Finance

Preview:

- A replacement for financial institutions.
- Preventing manipulation of data in centralized financial institutions.
- Solution to a repetition of data entry and paperwork.

"Finance is not merely about making money. It's about achieving our deep goals and protecting the fruits of our labor. It's about stewardship and, therefore, about achieving the good society."
- Robert J. Shiller

The world can change a great deal when the way the money moves and acts changes. This is why financial sectors are very keen on adopting emerging technology, that how the wheels which steer change is moved. As expected, the financial sector is adopting the blockchain technology before the other sectors. The concept of peer to peer transaction proposed by Satoshi Nakamoto was destined to change the financial system, the banks had the option of either getting toppled by this, or they could be a part of this change.

As compared to transaction methods like bitcoin, banks need a lot of revenue to function, and all the regulations are a terrible icing over the bitter cake. At first, regulations were on the side of banks trying to curb the industries getting crumbled, but soon enough financial institutions realized being the part of this change was inevitable. When cryptocurrencies like Bitcoin offers a hack free environment for money transactions which can stand tall when everything else is failing, it is an offer which is hard to refuse.

The blockchain technology being a permanent record will prevent any scams which can arise from manipulation of the documents. The entire list need not be visible to everyone, and the personal information can be encrypted so that it is visible only to suitable people. Unlike the currently centralized banks, the processes will be more transparent and reliable. The activities of every bank account will be easily verified and auditable, which will keep illegal activity under check. You can't even think of achieving some illegal feat now and manipulating the data later on.

The KYC implemented through blockchain will prevent unnecessarily and repeated paperwork, since the record can be shared with the concerned financial institutions very quickly and also the legitimacy of the documents can be easily confirmed. This will also ensure that activities like money laundering will be kept under check.

The transactions across borders will be made much more efficient with systems like ripple which provides faster and cheaper means of interbank transactions. With the world moving more towards digital money, any compromise on the centralized authorities which are managing it will create a big disaster. When the blockchain system is integrated with the idea of digital money, the safety of the system will be increased, and there will be more trust from the public.

The blockchain technology can also promote the micro investments by removing the minimum cap on investments. The transaction fee is significantly reduced with this system, which will remove the minimum cap from investment criteria. As discussed earlier, the systems like DAO will promote the growing companies and the ones requiring expansions by functioning like better versions of IPOs.

When micropayment systems are initiated, there will be the large door of opportunities open which will promote instantaneous transactions like you can pay for your electricity as you use it. With microtransactions in place, any undesired activities can be easily identified and kept in check by the machine learning process which will be continually looking out for anomalies in user activities. This doesn't mean that the credit system will vanish, we do not want all the opportunities that arise from the credit system to cut off.

Summary:
- Financial institutions are first to adopt emerging technologies
- Decentralized networks provide a means of cheaper and faster transactions.
- KYC procedures will ensure that unnecessary repetition of paperwork are avoided

7. Healthcare

Preview:
- Continuous evaluation of health and threat detection.
- Avoiding the data entry mess while switching healthcare institutions

"The art of medicine consists of amusing the patient while nature cures the disease."
-Voltaire

You don't want to always confined within the boundaries of the hospital if you want to monitor your health vigilantly. There are sensors available which can extract data like heart rate, pulse, glucose level as well as monitor sleep and other physical activities. When this data is communicated to the concerned doctor on a regular basis, post-treatment analysis or any irregularities can be taken off more efficiently. If there are irregularities in vitals, the doctor will be altered so that the issues can be taken care of immediately. The sensors which are available in the market now are exact, specific and sensitive. The data will be updated on a real-time basis, and during the patient's next visit, the doctors will have a better profile of the patient which will ensure better diagnosis.

A dedicated profile of the patient can be created in a portal which can be used for better patient-doctor communication, and just the sensor information won't be sufficient all the time. The activities like fixing suitable appointments will be more comfortable and efficient. An even mandatory hospital visit can be avoided with the help of this kind of profiles. The doctor can even provide a custom-made checklist which is to be updated by the patient on a regular basis - daily, weekly or monthly so that healthy habits are more enforced. Healthy food habits and dietary restrictions can be easily monitored using a combination of such checklist, the sensor information as well an option to upload pics along the checklist for confirmation. When the patient himself sees the effect of these habits from the favorable variations in readings of the sensors, they will be more motivated to stick to the plan and thus breeding healthier habits. It works just like the calorie tracking helps many people to stick to their weight loss plan, by seeing the more or less correlation of the calorie consumed to the amount of weight loss.

With the tremendous amount of data available from the sensors, machine learning can be implemented to keep track of the data and find the pattern continues. The patient can be alerted for a right diagnosis in case some abnormal patterns are detected. When more and more data is fed into the system, the efficiency of the machine learning can be further improved when the data is combined with the actual results of patients direct diagnosis. All this information related to a patient is being recorded and stored in the profile dedicated to the patient. The record used to create the profile will be immutable and secure through cryptography as and when required. The constant availability of real-time data can also improve the research and development in this sector. A network of experts can also be formed in this chain who can be called out for cross-reference of the diagnosis and confirmation of the analysis, which will increase the efficiency of the process. There is always a chance of human error when machine learning is applied human intervention is reduced to just overlooking the processes. This combination is a revolution in the decision-making process of the doctors.

With the implementation of blockchain in the sector, the cost of treatment will be significantly justified as the immutable data record stores everything related to the treatment of the patient. When a patient is visiting multiple hospitals, the immutable ledger can give a better insight into the patient's health history which will help the doctor profile the patient's health situation more efficiently. Each element used to diagnose and treat the patient will be tracked and detailed in the ledger which will create great transparency in the system. There will be no room for corruption by manipulating the process because there is always a necessity of the prices to be justified.

Summary:
- A universal ID for patients which can is encrypted and can be transferred to other healthcare institutions(Transparency in patient's health history).
- Sensors attached to the subject will update the vitals, any anomaly will be detected automatically, the concerned health specialist will be alarmed, and the appointment will be automatically confirmed.
- The cost of every treatment will be justified, with the entire lifecycle of treatment being recorded in an immutable ledger.

8. Insurance

Preview:
- Upgrading the current insurance system which is cumbersome and requires a lot of paperwork to process the claim

"There are worse things in life than death. Have you ever spent an evening with an insurance salesman?"
-Woody Allen

Let's imagine you are driving your car in future, well there is a possibility of cars being autonomous which does mean that there is no requirement for driving the car manually, but today you switched over to the manual mode because you felt so. When you switched to manual, the effective insurance policy gave you a notification that manual mode lets you claim only half the insurance amount since you were really in the mood for a drive, you confirmed the notification and started driving. After an hour or so of driving, you lost your attention for a second, and you ended up bumping into a pole and damaged the front of your car.

The sensors on your car picked up the impact details and immediately sends it to the insurance company. This also triggers the surveillance system to take pictures of the car so that the damage details are confirmed. You get a message that the insurance is approved well, of course, you receive only half the amount since you agreed to the manual drive clause before you started. In a matter of time, the replacement vehicle will arrive at the location and the damaged car will be sent to the repairing facility.

It is hard to digest how more comfortable these processes can become with the implementation of emerging technologies. Using smart contracts in blockchain network, the insurance premium can be transferred on a regular basis. All the automated responses will be initiated by the smart contracts like responding to the impact. The impact data is analyzed by the AI to understand the degree of damage, and it is correlated with the images obtained from surveillance. The smart contracts also ensure the confirmation of the fund, the process of sending the replacement vehicle and the tracking of repair progress. The safety of replacement vehicle can be determined by just checking the blockchain tag associated with it, even though it is sent after the automated verification, it can still be reconfirmed by scanning the blockchain tag.

The paperwork involved in current insurance processes are insane when this paper is supposed to pass over different departments the process will get very tedious and time-consuming. These hefty processes also call out for more transaction fees since a lot of manual effort is being put into it. The blockchain record is encrypted which can be shared with the other department just by providing access to the concerned department, and there won't be any data repetitions required to finalize the claim. The smart contracts will ensure that the conditions of the insurance are vigorously enforced.

In case of natural disasters, the IOT devices can be set to automatically dispense the amount like when there is the weather department can verify heavy seismic activity, this trigger from the sensors. The department will be included as a node in the blockchain record which will be updating details of natural calamities. When the details are updated or caught by the weather department sensors, the insurance amount will be dispensed to the people as soon as possible.

Summary
- The smart contracts will automatically trigger the claim based on the evidence collected from the sensors or the concerned departments.
- The confirmation paperwork can be avoided due to transparent and immutable nature of the ledger.

9. Law enforcement

Preview:
- The future of law enforcement

"It's about time law enforcement got as organized as organized crime."
-Rudy Giuliani

If you are aware of the movie "Minority Report" or the series "Person of interest," then you have a pretty good idea of how predictive policing works. Potential criminal activity is predicted or identifies using analytics, and this process is called predictive policing. If we are breaking down the prediction into different categories of how it can be prevented, it will be predicting the nature of the crime, predicting the potential victims and predicting identities of the criminals. The analytics can be designed to fit one or more of these categories. The concept of predictive policing is referred to as the process of stopping crimes before it happens and that's how the media is praising this concept.

In the George Orwell novel "Nineteen Eighty Four" introduced the term thoughtcrime which can be seen as the predecessor the idea of predictive policing and the term precrime was introduced in the Philip Dick short story called "The Minority Report", where Precrime is the agency which eliminates people who are going to commit crimes in future.

How is this accomplished?

Previous data is majorly used to find patterns and to correlate crimes to different parameters. When the locations are categorized geographically according to the risk associated with it, the police can distribute its force according to contributing maximum patrol over the areas with higher risks of crime. The law enforcement agents will more efficiently use the available resources with such kind of location tagging. This will enforce the idea of being at the right time at the right time.

According to LAPD when this was implemented, crime detection rates became double compared to previous methods and in Santa Cruz, the rate of burglaries dropped by about 20 percent when this was implemented over a period of 6 months.

Even though this looks promising, there are chances of the system to get biased based on how the baseline is set to detect crimes. One example is the possibility of racial profiling, which is a controversial topic when we are considering normal police behavior. According to Ezekiel Edwards, this is more of a predicting policing pattern than the predicting crimes which it claims to do.

The records and data which enables the prediction is susceptible and can't afford to be modified, so the solution is to design an encrypted, secure record. This data should not be readily available to everyone as it can cause chaos. Therefore the access should be well defined according to the hierarchy of the department. The answer for maintaining such a record is Blockchain which provides the transparency, security, and ease of auditing which is expected from such a system. The permanent nature of the blockchain network will ensure that the data is not manipulated.

The analysis is done by machine learning using the preexisting crime record data and by using iot sensor to update data regularly. Example of such sensors will be the ones who can detect emotional patterns by facial recognition in areas which are rated high in the pre-crime index.

When places of high crime probability are recognized, the places can be equipped with armed surveillance drones, which can keep better control of the situation. Armed drones are controversial because we do not want harmless people to be hurt therefore more R & D is required to make the armed drones function correctly from the machine learned data.

Other less harmful options include equipping the city with LED incapacitator or long-range acoustic devices which can be used to control people in a particular area. The bright lights from incapacitation cause nausea, vomiting and can even induce temporary blindness. These systems along with the Artificial Intelligence which processes and amends data in the permanent record can be an efficient law enforcing the system.

Summary
- The implementation of predictive policing which can forecast the occurrences of criminal activities and alert the concerned authorities.
- The working of this system greatly depends on the baseline. Therefore there are chances of the system to be biased.
- The sensitive records are maintained using blockchain which will prevent any manipulations

- Crowd control devices like LED incapacitation can be automatically deployed upon detection of criminal activity.

10. Real estate

Preview

- Why real estate should adapt emerging technologies
- Using virtual reality in selling real estate
- Blockchain and augmented reality for selling under construction plots
- Understanding user requirements

"Ninety percent of all millionaires become so through owning real estate. More money has been made in real estate than in all industrial investments combined. The wise young man or wage earner of today invests his money in real estate."
Andrew Carnegie

Real estate is responsible for a significant fraction of the world's wealth. What do you think the revenue model of McDonald's is? The real estate, the main reason which helped McDonald's to scale up was the adoption of the real estate based revenue model. The real estate sector is changing and adapting for the future. By learning how emerging technologies influence real estate, we can grab that sweet beginners advantage.

Most of the real estate deals happen on a direct basis through agents and the adoption of portals in real estate is decidedly less compared to the direct sales in the sector. While, looking for a property, our first resort is always to browse the availability on the internet, mostly by using our smart device. We prefer apps to check out listings, but the right 3rd parties like brokers do not efficiently use apps to interact with the buyers, more of a direct interaction comes into play.

Something which can boost the game of real estate listing application is the use of virtual reality. It is a big ordeal to show the clients each property, which is time and energy intensive process which can be replaced by a virtual reality preview, which will give a better idea and depth of the property before they can finalize to have a direct look. Even if the property is under construction, technologies like augmented reality can be used to show the final product. Along the proof of working timeline and goals, the pre-selling of such properties will be much more comfortable than showing them as images. The physical visit often requires the fixing of appointments which are very time consuming if there are multiple parties interested in the same property. This is also a highly third party dependent system with all the physical visit and the vibe provided by the agent will significantly determine the nature of the sale.

What if the data is manipulated? What if the augmented reality they are showcasing is an exaggerated version of the original products, like most of the advertisements, are doing now, giving us a big blow to our reality vs. expectation ride. Let's explore the possibilities of the internet of things, AI and blockchain in this sector before those questions are answered.

IOT is concerned with collecting data using sensors, how can this benefit the real estate? You can suggest the best property suitable for the clients using data of how much time they are spending in particular locations of the properties. Imagine you present the clients with VR version of 5 different properties, they give a negative response for 5 of those properties, because they did not find it satisfactory enough, but now you have the data of how much time spend in exploring different portions of the properties. With a great deal of data available regarding the time spent inspecting particular elements against chances of preferring can element can be derived using machine learning and artificial intelligence. This information can be gathered from all the 5 properties which were displayed; this will create a profile of preference for the client. Using this profile a property which matches the most with the taste of the client can be selected.

The data is pulled from a record of the real estate listing; blockchain can be implemented for managing this list. There are several elements involved concerning real estate which justifies its price like the location factor, the value added by the structure and the aesthetic value. When a permanent record is used to manage such a record, you can identify everything related to the property at a grass root level. You will feel satisfied with every penny you are spending on this property.

Let's get back to the case of data manipulation now when the property is backed by blockchain, the budget of the project involved in building the property will be linked to the elements involved in building, which leaves no chance for manipulation, forgery or any scams. For example, consider this situation, If the total budget of the property is $10 Million, the budget will be split and tagged to the respective resources using blockchain(tokenizing the budget will enable this process). If $0.50 Million is required for buying bricks in the property, that amount will be tokenized and will be tagged so that it can only be used to buy bricks. Every penny of the budget will be subjected to that kind of tagging, so this will ensure that whatever you see in your preview will be a real version of what you will be getting. The expectations will start meeting the reality of the implementation of the technological combo.

The requirement of brokers will be negated; there will only be people who are overseeing these more of an automated process. When a third party is eliminated, it means that the cost of the property will be reduced marginally. There will be a direct connection between the buyer and seller which is backed by transparency and security. The previously hassle-filled process will be much efficient and satisfactory now. All you have to do is browse through a bunch of virtual reality version and set a baseline for your preference and leave the rest of the work for the technology to do. Using this futuristic application, you will find your "Dream Home" which satisfies the title.

Summary
- Real estate is responsible for the significant fraction of the world's wealth
- Virtual reality and augmented reality can be for property tours without wasting time for appointments.
- IOT devices can be used to understand the preferences of the client so that satisfactory properties can be identified using machine learning.
- Blockchain will ensure the consistency of displayed data.

CONCLUSION

The Status quo bias

Have you experienced a queasy feeling to change the current state of things?

Let's take a situation, and you are given a ball to shoot through a hoop. You are supposed to attempt throwing the ball through the hoop 10 times. Imagine that you successfully passed the ball through hoop 5 times in those 10 attempts. Now you are offered a chocolate bar if you can score at least 5 in next 10 attempts and also you are offered two chocolate bars if you could score the same using your other hand, which is not dominant. In this situation, the chances are you opt for the first option. This choice is because of the regular use of your dominant hand in such activities which makes it more comfortable to attempt the feat using the dominant hand, and there is mental inertia involved if you try to switch the hands.

This phenomenon is known as Status quo bias, which makes us stick to the present state of things. Whenever any variations from this state happen, this is considered as a losing situation by our minds, and this is how we are programmed to react when something is shaking or when we are given an option to shake our baseline state.

The regret of losing is always more significant than the satisfaction of a win. It is, in fact, rational to stick to the status quo when the information provided regarded the topic is limited. When the information is limited, we can't quite understand the state in which we are given an option to switch to so practically it is wise to stick with current status quo. The decision is also viable when we would quite understand the situation, and this might be the case which requires a great deal of mental effort. Everyone's mind works in a different spectrum, and we can't expect everyone to deduce everything therefore in this case we can opt to maintain the status quo.

Imagine you are trying to find a partner when you finally get someone, you may have this feeling in your head that there is a chance that you might get someone better and you should keep looking. This feeling is mental inertia of keeping on searching for better things in action. Your fear of making the wrong decision increases the chances of you sticking to the status quo.

When you are frequently exposed more to a particular situation, it is perceived as a baseline status quo. The familiarity factor drives the feeling that the relevance of the situation is making it appear over and over again.

You change your status quo if there is an equal chance of a win and lose. The magnitude of satisfaction caused by the winner is less than the magnitude of regret caused by the loss. This feeling makes us stick to the status quo, and this thought process is known as loss aversion.

What can you do to be not influenced by the status quo bias and to take the best decision possible considering all the available resources?

The solution is to reverse the thought process. You should assume that the end state is your status quo and you should assume your current status quo as an end state. After visualizing this situation, if you still find the 2nd state to be evil like you did when the bias influences you. If the answer gives the positive mental outcome in one direction and gives negative one in the other, then you successfully fought off the bias, and you should go for the direction which suggests the positive outcome

The status quo bias can be seen in action when the adoption of emerging technologies is involved. People do not want to change the present state of things; they are used how the industries are working. The people are used to services they create and the services they used. Status quo bias is a big wall which is to be broken to adopt new technology. With the basic knowledge of Blockchain, AI and IOT let's put our knowledge into action and break those walls.

Congratulation!

Congrats, you have successfully averted an opportunity lose. Now time does the rest of the job. If you find this promising and if you have a hunch that this could make you millions in future or even revolutionize the world, you will for sure start learning and strive to materialize these concepts. This awareness is not an end of your emerging technology journey, and it is the start of your new life. Maybe you felt like you just broke the cocoon and is ready to fly into endless possibilities now.

This read is just a start of your big journey; you get a chance to continue the journey with us by staying updated with everything related to these emerging technology in our website coachblink.com, which will actively guide you through the journey of revolutionizing the world. You can feel free to bring forward new ideas and possibilities of the emerging technologies. Subscribe to our newsletter so that you won't be missing out a single update on the website. Remember to keep this book within your reach so that you are frequently reminded about the spark generated in you for changing the world

If you did not find this interesting enough to explore, maybe you should keep this around because there is nothing more promising like emerging technologies. There is no pressure to read, keep this around within your reach. Do not let this vanish into your collection of eBooks. When the time is right, you will pick it up again and it another try, and it leaves you wondering why this spark did not come in the first and maybe you realize you did not give it enough chance. After all, the chance is all we have to give, and the chance is all it takes to make a change.

Don't forget to keep yourself updated at the official coachblink website
www.coachblink.com

Subscribe to our newsletter using your email ID. Stay tuned for books and other stuff which can potentially change your life.

In case of any technical queries and requirements please drop us a mail to
info@coachblink.com.

Disclaimer:

This book doesn't provide any legal or financial advice regarding cryptocurrencies and related applications.

www.ingramcontent.com/pod-product-compliance
Lightning Source LLC
Chambersburg PA
CBHW030943240526
45463CB00016B/1447